Diamonds in the Rough

NEW AFRICAN HISTORIES

SERIES EDITORS: JEAN ALLMAN, ALLEN ISAACMAN, AND DEREK R. PETERSON

*Books in this series are published with support from the
Ohio University National Resource Center for African Studies.*

David William Cohen and E. S. Atieno Odhiambo, *The Risks of Knowledge: Investigations into the Death of the Hon. Minister John Robert Ouko in Kenya, 1990*

Belinda Bozzoli, *Theatres of Struggle and the End of Apartheid*

Gary Kynoch, *We Are Fighting the World: A History of the Marashea Gangs in South Africa, 1947–1999*

Stephanie Newell, *The Forger's Tale: The Search for Odeziaku*

Jacob A. Tropp, *Natures of Colonial Change: Environmental Relations in the Making of the Transkei*

Jan Bender Shetler, *Imagining Serengeti: A History of Landscape Memory in Tanzania from Earliest Times to the Present*

Cheikh Anta Babou, *Fighting the Greater Jihad: Amadu Bamba and the Founding of the Muridiyya in Senegal, 1853–1913*

Marc Epprecht, *Heterosexual Africa? The History of an Idea from the Age of Exploration to the Age of AIDS*

Marissa J. Moorman, *Intonations: A Social History of Music and Nation in Luanda, Angola, from 1945 to Recent Times*

Karen E. Flint, *Healing Traditions: African Medicine, Cultural Exchange, and Competition in South Africa, 1820–1948*

Derek R. Peterson and Giacomo Macola, editors, *Recasting the Past: History Writing and Political Work in Modern Africa*

Moses E. Ochonu, *Colonial Meltdown: Northern Nigeria in the Great Depression*

Emily S. Burrill, Richard L. Roberts, and Elizabeth Thornberry, editors, *Domestic Violence and the Law in Colonial and Postcolonial Africa*

Daniel R. Magaziner, *The Law and the Prophets: Black Consciousness in South Africa, 1968–1977*

Emily Lynn Osborn, *Our New Husbands Are Here: Households, Gender, and Politics in a West African State from the Slave Trade to Colonial Rule*

Robert Trent Vinson, *The Americans Are Coming! Dreams of African American Liberation in Segregationist South Africa*

James R. Brennan, *Taifa: Making Nation and Race in Urban Tanzania*

Benjamin N. Lawrance and Richard L. Roberts, editors, *Trafficking in Slavery's Wake: Law and the Experience of Women and Children*

David M. Gordon, *Invisible Agents: Spirits in a Central African History*

Allen F. Isaacman and Barbara S. Isaacman, *Dams, Displacement, and the Delusion of Development: Cahora Bassa and Its Legacies in Mozambique, 1965–2007*

Stephanie Newell, *The Power to Name: A History of Anonymity in Colonial West Africa*

Gibril R. Cole, *The Krio of West Africa: Islam, Culture, Creolization, and Colonialism in the Nineteenth Century*

Matthew M. Heaton, *Black Skin, White Coats: Nigerian Psychiatrists, Decolonization, and the Globalization of Psychiatry*

Meredith Terretta, *Nation of Outlaws, State of Violence: Nationalism, Grassfields Tradition, and State Building in Cameroon*

Paolo Israel, *In Step with the Times: Mapiko Masquerades of Mozambique*

Michelle R. Moyd, *Violent Intermediaries: African Soldiers, Conquest, and Everyday Colonialism in German East Africa*

Abosede A. George, *Making Modern Girls: A History of Girlhood, Labor, and Social Development in Colonial Lagos*

Alicia C. Decker, *In Idi Amin's Shadow: Women, Gender, and Militarism in Uganda*

Rachel Jean-Baptiste, *Conjugal Rights: Marriage, Sexuality, and Urban Life in Colonial Libreville, Gabon*

Shobana Shankar, *Who Shall Enter Paradise? Christian Origins in Muslim Northern Nigeria, ca. 1890–1975*

Emily S. Burrill, *States of Marriage: Gender, Justice, and Rights in Colonial Mali*

Todd Cleveland, *Diamonds in the Rough: Corporate Paternalism and African Professionalism on the Mines of Colonial Angola, 1917–1975*

Diamonds in the Rough

*Corporate Paternalism and African
Professionalism on the Mines of Colonial
Angola, 1917–1975*

～

Todd Cleveland

OHIO UNIVERSITY PRESS ～ ATHENS

Ohio University Press, Athens, Ohio 45701
ohioswallow.com

© 2015 by Ohio University Press

To obtain permission to quote, reprint, or otherwise reproduce or distribute
material from Ohio University Press publications, please contact our rights
and permissions department at (740) 593-1154 or (740) 593-4536 (fax).

Printed in the United States of America
Ohio University Press books are printed on acid-free paper ∞ ™

25 24 23 22 21 20 19 18 17 16 15   5 4 3 2 1

Library of Congress Cataloging-in-Publication Data
Cleveland, Todd, author.
   Diamonds in the rough : corporate paternalism and African
professionalism on the mines of colonial Angola, 1917–1975 / Todd Cleveland.
      pages cm. — (New African histories)
   ISBN 978-0-8214-2135-2 (hc : alk. paper) — ISBN 978-0-8214-2134-5
   (pb : alk. paper) — ISBN 978-0-8214-4521-1 (pdf)
   1. Diamond industry and trade—Angola.   2. Diamond mines
and mining—Angola.   3. Diamond miners—Angola—Social
conditions.   I. Title.   II. Series: New African histories series.
   HD9677.A542C55 2015
   338.2'78209673—dc23
   2015004834

To Julianna

Contents

Illustrations

FIGURES

MAP

TABLES

Acknowledgments

This project would never have progressed much beyond the conceptual stage without the assistance of a great number of people, spanning three continents. First I'd like to thank Patricia Hayes, who introduced me to the topics that this book explores. Following a talk she gave at the University of Minnesota, her comment that a study of diamond mining in colonial Angola would constitute a long overdue endeavor almost instantaneously concluded my struggle to identify an appealing project. From the commencement of the research to the completion of the book, Allen Isaacman has been intimately involved; his ongoing support and encouragement throughout have been unflagging. At Minnesota, I also received indispensable training, support, advice, and feedback from several other colleagues, including, most notably, Helena Pohlandt-McCormick and Fernando Arenas. Before arriving at Minnesota, I received invaluable assistance from Doug Wheeler and Funso Afolayan at the University of New Hampshire.

Research in Portugal could not have been completed without the assistance of several individuals. In Lisbon, Franz-Wilhelm Heimer and Ana Paula Tavares offered advice and maximized my time and efforts by directing me to the most instructive archival sources. Staff at the Arquivo Histórico Ultramarino, Biblioteca Nacional, Sociedade de Geographia, and Torre do Tombo were always welcoming and helpful. In particular, Aura Carrilho made my time at the Torre do Tombo as productive as possible, and always did so with a smile, while Isabel Reis introduced my wife, Julianna, and me to the extensive network of former Diamang employees residing in Portugal, and in all other ways treated us as if we were lifelong friends.

In Coimbra, Nuno Porto's contributions were immeasurable. As overseer of the Anthropology Department's Museum at the Universidade de Coimbra, Nuno facilitated access to Diamang's archive, which is housed at the museum. When Nuno first opened the shed in which

Diamang's uncatalogued records were stored, I experienced a moment about which every researcher dreams. During the ensuing six months or so that I waded through this archival treasure trove, Nuno arranged for daily access to the materials and the use of an office. He also shared with me his significant knowledge of Diamang, which I repeatedly tapped to answer the questions that I was formulating about the company and its African labor force.

It was also in Coimbra that I first met my great friend Jorge Varanda. As someone who had been interested in Diamang's Health Services for years by the time I started my project, Jorge patiently endured my efforts to "catch up." Although we focus on different aspects of the diamond enterprise's history, our significant overlapping interests led to fieldwork together in Angola and a seemingly endless exchange of ideas and feedback concerning our respective Diamang-related efforts. Jorge's knowledge and analysis of Diamang's operations inform this book at every juncture.

In Angola, logistical problems were myriad and unrelenting, but owing to the assistance of key individuals, were ultimately overcome. From the moment we arrived, Julie Thompson and Roquinaldo Ferreira provided invaluable support as we attempted to negotiate and navigate postconflict Angola. Rob Miller, Christine Wilkins, Jamie Fisfis, Ben Osland, Isabel Emerson, and Ranca Tuba, staff members at the joint offices of the National Democratic Institute and the International Republican Institute in Luanda, were uncommonly generous. The staff at the Arquivo Histórico Nacional in Luanda was always patient and helpful. When I wasn't in the field, I spent my days plodding through boxes (*caixas*) of colonial-era records, which staff members tirelessly retrieved from dimly lit recesses. The director of the archive, Dra. Rosa Cruz e Silva, ensured access to these materials throughout my stay and also furnished the crucial "letters of invitation" that enabled me to procure a series of visas required to (re-)enter the country.

Without significant logistical support in Angola's Lunda Norte Province, Diamang's former area of operations, this project would be incomplete, at best. In order to travel to and within this turbulent corner of Angola, I depended on a number of individuals. In and around the provincial capital, Dundo, Padre Damião was a generous and helpful host. Provincial Governor Francisco Gomes Maiato also supported my work, eventually offering me quarters in the official state residence ("K-18")—formerly the accommodations of Diamang's director in

Angola. While I was a guest there, Paulina Lassalete helped with translations and always made me feel at home. Pedro Blayr helped arrange interviews, answered my far-ranging questions, and welcomed me into his home. Mick Comerford and Francisco Terra also played key roles in facilitating my trips to Dundo.

In the western stretches of Lunda Norte province, I received priceless support from ITM Mining. Employees Andy Machin and Carl Niemann arranged for me to stay with the company in its compound along the Cuango River, where they housed and fed me and permitted me to observe diamond mining operations firsthand. ITM staff also provided security and transport and eventually introduced me to local residents who both organized many of my interviews and helped with translations.

Alberto Baião, Caiombo Jombe, José Turiambe Muachiriango, Paciencia Roberto António Constantino, and Zach Manuel were among those individuals who tirelessly assisted me with the interview process in western Lunda Norte. For days on end, they voluntarily translated Chokwe to Portuguese and vice versa, often during blisteringly hot extended interview sessions. They also helped arrange these interviews, locating a variety of informants who could provide varying perspectives on life at Diamang.

These informants arguably warrant the most gratitude. They unflinchingly shared the most intimate details of their Diamang experiences with a perfect stranger and exhibited the utmost patience when fielding questions that I later realized were based on inaccurate or even absurd assumptions. Given the difficulty of surviving in contemporary Lunda Norte, their willingness to share their time—often waiting for days to secure an audience with me—underscores both their generosity and their courage.

Facilitating each step of the research process was the abundant funding I received in support of this project. A series of US Government Foreign Language and Area Studies Fellowships awarded by the University of Minnesota enabled me to build my Portuguese language skills over the course of two summers and a full academic year. Travel money from the MacArthur Program and History Department at Minnesota facilitated travel to Portugal and Angola, providing essential support for my initial exploratory visits to these sites. An international research grant from the Graduate School at the University of Minnesota and additional funding from MacArthur helped us establish ourselves in Portugal for six months before proceeding on to Angola. Once in Angola, where the cost of living is often mind-boggling, funding from the Fundação Calouste Gulbenkian and the US Department of Education, in the form of a Fulbright-Hays Research Fellowship, proved vital. Following two continuous years of fieldwork in Angola, we relocated to Portugal, where financial support from the Council for European Studies at Columbia University and a fellowship at the Torre do Tombo, both funded by the Luso-American Development Fund, enabled me to satisfactorily conclude my research. Marta Abrantes also offered important assistance during this period.

On returning to the United States, I served as a fellow at the Carter G. Woodson Institute at the University of Virginia. It was at this point that I began the daunting task of molding a mountain of archival and oral research into something much more digestible and, ideally, compelling. During my years at Virginia, Joe Miller offered unremitting support, helping me through the formidable intellectual and personal challenges that the writing process inherently generates. The friendship that grew out of these and other interactions endures.

When the time came to prepare the final version of this book, Gill Berchowitz at Ohio University Press, as well as Allen Isaacman, Jean Allman, and Derek R. Peterson, the editors of the series in which this volume appears, provided sage advice and consistent support. Anonymous readers similarly offered useful feedback and comments that led to the production of a tighter, more focused narrative.

It was during this extended period of revision that Dave and Suzie Parkinson—whose cousin and father, respectively, Lute Parkinson, had served on the Angolan mines during the early decades of operations—generously sent me a series of insightful materials, including Lute's diary chronicling his time in Africa.

Throughout this protracted process of research, writing, and revision, I enjoyed the steadfast support of two families: my own and Julianna's. My parents, Roger and Karen, and my brother, Brad, encouraged me at every step. And, as my sister Kim's academic career roughly parallels my own, she supplied invaluable intellectual support, as well as commiseration and humor at all the right times. Julianna's parents, Wally and Joyce Munden, deserve special mention for their comprehensive support. Their assistance was nothing short of remarkable; without it, my lengthy fieldwork experience would not have been feasible.

Finally, and most importantly, I would like to thank Julianna. In Portugal, she acted as a research assistant—among countless other contributions—while her hard work at IOM in Angola provided crucial funds for both essentials and frivolities in that exorbitantly expensive milieu. Her indefatigable optimism carried me through my deepest moments of "research despair," and her unwavering companionship unfailingly spurred me onward as I emerged from these hollows. I can only hope that our two sons will inherit these qualities from their extraordinary mother. Muito Obrigado, Julianna.

1 ⟿ An Introduction to Angola's Diamond Past

Paternalism, Professionalism, and Place

> I admired the perfect organization of Diamang, I admired its industrial
> activity, I admired . . . the riches of its mines. But, more than the riches of
> the diamonds, I admired the richness of its souls.
>
> —Dr. Vieira Machado, Portuguese Minister of the Colonies,
> following a 1938 visit to Diamang's installations

> Diamang has to be appreciated and judged not only . . . as a mining
> company, that extracts and sells diamonds. . . . This enterprise, as a
> consequence of the unique conditions in which it was created, and . . .
> from its isolation in relation to the rest of the colony . . . has evolved to be
> more an "enterprise of colonization" than a simple mining undertaking.
>
> —Diamang General Assembly Meeting, 1959

ON THE MORNING OF November 4, 1912, prospectors found seven
small diamonds near Musalala Creek in the Lunda region of the Portuguese colony of Angola. Less than five years later, in October 1917, international investors formed the Companhia de Diamantes de Angola,
or Diamond Company of Angola (Diamang), to exploit the alluvial
diamond deposits that had been identified in the interim.[1] By 1921,
in exchange for the rights to half of the company's annual profits, the
Portuguese colonial government had granted Diamang exclusive mining and labor procurement rights over a vast concessionary area in the
northeastern region of the colony, roughly the size of the state of Okla-
homa (map 1.1). Using these monopolies, the self-described "enterprise
of colonization" was to become the largest commercial operator and

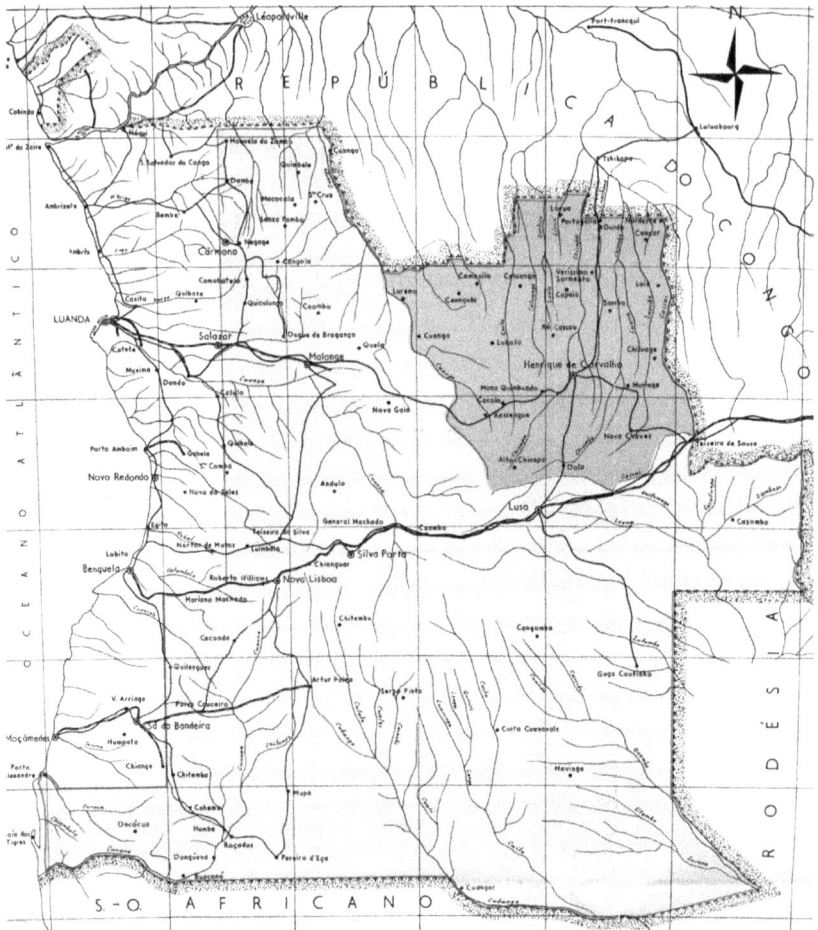

MAP 1.1. Angola, c. 1963, featuring Diamang's expansive mining concessionary area (in lighter shading) and its smaller operational and exclusive labor procurement zone in Lunda (in darker shading). *Source: Science Museum of University of Coimbra.*

leading revenue generator in the history of Portugal's durable empire in Africa. Only the protracted Angolan Civil War (1975–2002), which followed the country's independence from Portugal in 1975, was able to finally bring this industrial leviathan to its knees.[2]

It was on the backs of Diamang's African labor force that the diamond enterprise generated its prodigious profits. Many local residents sought employment with the company, though others were forcibly recruited from throughout the Lunda region and brought to work on the mines. Together, they constituted the collection of "rich souls" that so impressed Dr. Vieira Machado following his visit to the company's installations in 1938, as noted in the epigraph. Because Diamang demonstrated a preference for manpower over more costly machinery throughout its history, it aggressively pursued ever-greater numbers of laborers to staff its expanding operations. Consequently, from the commencement of mining operations in 1917 until Angolan independence in 1975, approximately one million African men, women, and children—who often traveled to the mines together as families—toiled for Diamang. Based on the company's ability to procure and service this massive labor force, its operational zone became known in both European and African popular imaginations as *um estado dentro do estado*: "a state within the state."

Following a tumultuous initial decade and a half of operations, a remarkable level of stability pervaded the company's mines and mining encampments, as well as the thousands of kilometers of countryside that surrounded these installations. This relative quiescence stands in sharp contrast to the labor strikes, trade unionism, and intra-mine ethnic conflict that African mine workers elsewhere so commonly generated. Lunda remained conspicuously quiet even during the Angolan War for Independence (1961–75), while much of the rest of the colony erupted in violence. The central question that this book attempts to answer is: Why, in light of the demanding labor regime in Lunda, did African mine workers not adopt a more militant posture?

One way to interpret the absence of unrest on and around the mining installations is to attribute it to the repressive capabilities of Diamang and the colonial state. After all, the company was the largest and most powerful in Portugal's African empire. It was, therefore, able to stipulate that the state, which often projected terror to compensate for severe human and material resource shortages, support its operational objectives.[3] However, even after a variety of international pressures in the early 1960s—more than a decade before the conclusion of the company's mining operations—compelled Diamang and the state to abandon corporal punishment and the forced labor scheme, stability reigned in Lunda. In fact, an inverse relationship existed between violence and productivity on Diamang's mines: as the former decreased, the latter increased. While acknowledging that both real and potential violence contributed to keeping the regional population subdued—both on and off the mines—this book argues that a unique blend of company pragmatism, paternalism, and profits; African workers' occupational and social professionalism; and Lunda's geographical isolation were primarily responsible for the exceptional quietude.

It was during the 1930s that the confluence of these factors first began to produce this stability. Prior to this decade, many Lunda residents had violently resisted the company's presence or fled ahead of labor recruiters. Similarly, Diamang's African employees often deserted, while others only halfheartedly worked if they stayed. For its part, the company was offering only rudimentary accommodations, insufficient rations, and low wages, and mine overseers regularly assaulted laborers. However, as Diamang's operations grew, company officials realized that they desperately needed the labor latent in the scarcely populated region. Moreover, as Diamang was unable to attract significant numbers of workers from beyond Lunda, its African employees were not expendable the way that they were on mines elsewhere on the continent and, in many cases, needed to be "recycled." In response, Diamang adopted a pragmatic approach to its manpower needs. As part of this multifaceted strategy, the company improved working conditions, barred traditional authorities, or *sobas*, from its mines, and positioned Diamang to assume the paternalistic role of a "big man," with all of the reciprocal structures and tropes with which Africans were familiar. Testimony from Joaquim Trinidade, a former employee, captures the company's approach: "Diamang exploited the soil here, but also treated us [workers] well. So, Diamang was taking with one hand, but giving back with the other."[4]

The origins of this "giving back" can be traced to the 1930s—paradoxically, at the height of the Great Depression. Although the global economy had collapsed, and the worldwide demand for diamonds had correspondingly plummeted, De Beers's previously negotiated agreement to purchase Diamang's output to maintain its (near) monopoly on rough stones continued to buoy the Angolan company's revenues. While the global economy limped along, Diamang's annual sales almost *tripled* over the course of the 1930s. Meanwhile, De Beers was forced to halt production and shutter diamond mines in South Africa and neighboring South West Africa (Namibia) during this same period. This extraordinary scenario is especially significant because the 1930s was the most pivotal, transformative decade in Diamang's history, a period in which the company invested heavily in its health, food, and human infrastructures; as much of the continent suffered under the weight of the Depression, working conditions at Diamang were better than they'd ever been. If this newfound paternalism was rooted in the company's pragmatic approach to staffing challenges, these ameliorative initiatives were facilitated by the ever-escalating revenues that it was enjoying. Indeed, it is almost inconceivable that Diamang was so profitable—growing output, revenues, and the size of its labor force year after year—in the context of such severe global economic devastation.

Exempt from all state taxes and duties and bolstered by the lucrativeness of Lunda's high percentage of gem-quality stones (upwards of 80 percent), Diamang proceeded on very solid financial footing. From the 1930s on, it began using a portion of its wealth to upgrade services for its African workforce, including dramatic improvements in housing and health care; the introduction of a wide range of recreational activities; and an aggressive campaign to achieve full food security. Diamang also began dependably honoring the lengths of African laborers' contracts, and mine overseers administered corporal punishment increasingly selectively. Collectively, these calculated measures proved to be highly efficacious, as each year the number of voluntary workers grew, and the company enjoyed uninterrupted increases in annual revenues.

Absent from this battery of corporate improvements were elevated wages. Over time, Diamang only reluctantly and minimally raised wages, such that, for example, in the 1950s its rates were among the lowest in the colony. Given its regional monopoly on labor, competitive wages were simply unnecessary. Consistent with its paternalistic approach, Diamang instead allocated funds to enhance the overall well-being of both its African workforce and the regional population via an array of health care initiatives and improved nourishment; for company officials, a healthy workforce (and labor pool) constituted a productive workforce. The combination of low wages and benefits aimed exclusively at regional residents explains Diamang's inability to attract laborers from beyond Lunda. Low levels of monetary remuneration also ensured that minimal cash circulated in the region, money that in most other colonial mining contexts generated severe social disruption. By design, Diamang was to be the sole provider in Lunda.

Meanwhile, as the diamond enterprise consolidated its regional hegemony, Lunda residents-cum-mine workers determined that they, too, were "stuck" in a partnership with Diamang. In a region previously ravaged by the slave trade and most recently devastated and further depopulated by a collapse in rubber prices, sobas were no longer able to provide in ways that the company now could.[5] In turn, regional residents increasingly began engaging with Diamang in a manner that signaled that they had conceded to the company the paternalistic "big man" role it had been seeking to assume. For example, as Diamang began improving conditions on its mines during the 1930s, African laborers' productivity increased, while their absenteeism and desertion rates plummeted. I understand this form of strategic reciprocation as *occupational professionalism*, which stressed commitment and cooperation, while largely eschewing confrontation and subterfuge. More specifically, the constituent actions of this novel work ethic, which eventually became normative, included arriving to work on time; dutifully completing daily tasks; abstaining from work slowdowns, strikes, or other disruptive activities that would jeopardize production—the hallmark of mine laborers across much of Africa; and cooperating with co-workers across an array of potential social divides.

Testimony from Rodrigues, who began at Diamang in 1958, captures African laborers' promotion and application of professionalism: "On our way to the mines, we had plenty of opportunities to talk and mingle with workers who had just finished. These workers provided us with both information and advice. They told us to work with *força* [effort]—if you had two meters of gravel to remove, then do it with força. And we did."[6] Women at Diamang also cultivated and embraced this occupational approach. For example, Mawassa Mwaninga, who first ventured to the mines with her husband in 1964, offered the following testimony. "I had a baby girl while at the mines, in a company hospital. The *branco* [Portuguese] told me I didn't have to work while I was pregnant, but I did anyway—all the way up until five days before I gave birth! I received a week off afterwards, but was still being paid. After this, I went back to work with the child on my back."[7] Both Rodrigues's and Mwaninga's words powerfully convey the occupational commitment and focus that the enterprise's African laborers so consistently displayed.

As there had been no colonial precedents in Lunda before Diamang's arrival, it took workers time to formulate new conceptions of power, responsibility, and expectation in the context of labor-management relations on the mines. As the company began cultivating a more agreeable environment in the 1930s, African laborers began infusing the Western (corporate) notions of time and work that they had been internalizing with their own notions of social reciprocity. These complementary developments enabled workers to realize both their personal expectations and professional objectives.

Beyond forging a particular occupational approach, African laborers also adopted an assortment of social improvement strategies away from the work site, which were, in almost all cases, not intended to undermine Diamang's bottom line. For example, family members regularly accompanied recruits to the mines and, once there, creatively distributed workloads among themselves and reached across a range of social divides to befriend and fraternize with fellow residents in mine encampments, as well as with members of nearby communities who were otherwise unaffiliated with Diamang. I understand these actions and behaviors to be constitutive of *social professionalism*. Collectively, the dual social and occupational strategies that workers adopted enabled them to complete labor contracts without incident and, thus, return home to resume their pre-Diamang lives. Although company officials regularly lamented these departures, they also recognized how critical the local reproduction of the labor force was.

Although the African workforce and company officials were instrumental in maintaining stability in Lunda, the geographic isolation of the mines is analytically salient as well. Located in Angola's northeastern corner, Lunda was (and remains) a highly remote and scarcely populated area that initially presented innumerable problems for Diamang, including a lack of infrastructure and, therefore, desperately long and circuitous supply lines. However, this isolation eventually worked in the company's favor. By 1930, Diamang had established a security perimeter that allowed it to contain movement into and out of the region; in less remote parts of the colony, such a circumscription would have been futile.[8] This approach dovetailed with the company's development of a labor regime that featured close monitoring to prevent diamond theft, but lacked the brutality that often prompted it. Lunda's isolation also enabled Diamang to control what little urbanization did occur within its installations. Elsewhere in Southern Africa, the formal and extralegal remunerative options generated by urban development near mining operations profoundly unsettled those settings. However, even Dundo, the site of Diamang's headquarters in Lunda, was a modest and highly regulated space, which counted only a few thousand residents. In practice, Lunda's isolation compelled the company to carefully cultivate its relationship with the local population.

Collectively, African workers' social and occupational professionalism, Diamang's repressive capabilities and pragmatic, profit-fueled paternalism, and Lunda's isolation engendered a remarkable, perhaps unparalleled, degree of stability. Committed to completing daily tasks—though rarely to the company itself—Diamang's African employees identified a stable environment as the most desirable in which to satisfy their contracts and return to their home villages. Consequently, I contend that African adult and child laborers, rather than being victims of a violently generated and enforced quietude, actively and "professionally" participated in the process of stabilizing Lunda.

This book explores the extraordinary relationship that Diamang and its African workforce co-cultivated through an examination of the daily lives and experiences of these laborers over the course of roughly sixty years (1917–75). I examine the shifting strategies that shaped company–worker interactions as recruits and family members traveled from their villages to the mines, toiled on and around them, and eventually returned home. By reconciling the harshness of the regional labor regime with the relatively agreeable conditions and attendant dispositions of workers on Angola's diamond mines, the book strives to prompt new ways of thinking about how Africans in colonial contexts engaged with forced labor, mining capital, and, ultimately, each other.

A LOCAL MINING COMPANY WITH GLOBAL DIMENSIONS

Although Diamang featured a Portuguese veneer and its mining operations were confined to Lunda, the company was funded by private and corporate investors from around the world, and its capital and profits circulated across a network of global financial centers. The international dimensions of the enterprise, combined with the increasing power and autonomy vis-à-vis the colonial government that its profits were facilitating, deepened impressions of Diamang's operational zone as "a state within the state." Within this insulated area, the company's cosmopolitan character influenced its operational ethos and thereby shaped the lived experiences of its African workforce. For example, in Diamang's initial decades, members of its (white) managerial and engineering staffs moved regularly between Europe, North America, and other places in Africa, namely the Belgian Congo and South Africa, bringing to Angola not only their technical skills, but also empirically derived ideas of how to organize and discipline labor. Only by the 1940s, had a gradual process of "Portuguesization" largely deprived the staff of its global character, though even then, these new caretakers upheld the company's tradition of diversity in managerial philosophy.

Diamang's progressive operational area is also indicative of the outward-looking, international nature of the company; during the colonial period, Angola's capital city of Luanda constituted the only comparable space. Within Lunda, the diamond enterprise constructed a hydroelectric plant, an airport, an extensive road network, the most advanced health care infrastructure in the colony, and a museum that served as a global research center, hosting scholars, scientists, and doctors from around the world. Diamang officials also regularly interacted with international leaders in mining and medicine and imported this expertise and associated technologies to its domain, bypassing the rest of the colony.[9] In fact, because of the company's significant regional investments, this former "enclave" in Lunda remains one of the most developed spaces in contemporary Angola.[10]

THE UNIQUENESS AND BROADER HISTORIOGRAPHICAL SIGNIFICANCE OF DIAMANG

This study builds on and extends the rich literature on mining experiences in central and southern Africa, most of which have focused on South Africa, and to a lesser extent on Zimbabwe and Zambia. The history of Angola's diamond mines also departs in several significant ways from the histories of mining operations and experiences elsewhere on the continent. The book's title, *Diamonds in the Rough*, alludes to some of these core divergences. First, Diamang constitutes a type of corporate "diamond in the rough," considering its exceptional approach toward its African workforce, which, although undeniably exploitative, was not nearly as harsh as labor regimes on mines elsewhere in colonial or South Africa. This discrepancy prompts a reconsideration of the durable notions of labor in Portugal's African colonies. Second, the African workforce's reciprocal professionalism also renders it a "diamond," when compared with the more hostile, restive, or "rough" African labor forces that featured on mines throughout the continent. Finally, given the location of the mining operations—deep in the interior and well removed from other population centers, coupled with the considerable social and material development that Diamang realized in Lunda—the region constituted a type of "radiant" space in the otherwise unforgiving, bleak colonial Angolan "bush."

In the following section, I focus on six concrete aspects of Diamang's history that highlight its uniqueness and also the broader historiographical and analytical utility that these exceptionalities offer. These dimensions include the extended nature of the labor process; workers' strategic professionalism; the durability and impact of the forced labor regime; the presence and formal contributions of women and children; the paucity of internecine violence in mining encampments; and the development of a virtually autonomous "state within the state." Through an analysis of this array of distinguishing features, the study opens up new ways of thinking in these key areas.

First, the book expands conventional spatial and temporal notions of the labor process by defining it as the work performed en route to the mines; during the daily work regimen once on the mines; and also each evening "after the whistle blew," which consisted of the range of domestic tasks carried out in company encampments. Patrick Harries's examination of Mozambican mine workers constitutes a notable exception to the traditional historiographical focus on mine work sites and encampments. However, Harries's study explores these mine workers' cultural transformations as they trekked between South Africa and colonial Mozambique, rather than the labor performed during these passages.[11] By situating the beginning of the labor process at the point of recruitment, this book extends the analytical notion of "work" beyond the mines themselves to the areas from which laborers were drawn. This analytical broadening facilitates the inclusion of workers' journeys to and from the mines, which could take weeks, or even months, to complete (until the 1940s, when Diamang introduced truck transport to expedite this process). Meanwhile, the inclusion in the labor process of work performed in mine encampments each night, such as retrieving water and preparing meals, highlights women's and children's contributions and experiences.

In fact, the mere presence of women and, at times, children in Diamang's encampments renders Angola's mines different from those found elsewhere in colonial Africa. The Copperbelt that stretched across Zambia (Northern Rhodesia) and the (Belgian) Congo constitutes the sole exception, while remaining only partially analogous. Studies exploring mining scenarios in that milieu demonstrate that, like Diamang, certain mining companies encouraged laborers to relocate to the work sites with their families, even if remunerated opportunities for women on the Copperbelt were relatively limited.[12] This study builds on this literature, and, in particular, notions that workers accompanied by their wives had lower rates of morbidity, volatility, and absenteeism. However, I focus more squarely on the range of women's salaried and domestic tasks at Diamang and how their efforts enhanced their husbands' productivity. Company officials astutely recognized that to maximize profits they had to create an environment in which spousal accompaniment would be agreeable, if not attractive. To this end, the company provided superior housing for married couples and families, as well as clothing, health care, and a range of formal, remunerative opportunities for women.

Second, the book introduces the dual concepts of social and occupational professionalism to reflect and explicate African laborers' studied, deliberate patterns of reciprocal behavior, developed in response to Diamang's ameliorative initiatives. Heretofore, scholars of Africa have rarely invoked "professionalism," reserving it for studies of indigenous soldiers.[13] Instead, this book understands professionalism as the willing demonstration of employer-defined appropriate conduct and explores how African laborers' strategic adoption of and adherence to this conduct served both their remunerative and social objectives, while also contributing to regional stability.

Scholars of Africa's colonial past have long sought to identify instances of worker resistance, searching for ever-more-subtle, creative forms. When absent, "collaboration" or prohibitively extreme oppression is typically cited. I, too, approached Diamang's history in search of African workers' acts of resistance but failed to encounter any sustained, aggressive examples. Yet I quickly concluded that the usual explanations did not apply; the demanding conditions on the mines seemed neither determinative nor preclusive of a more measured response by the African labor force. Rather than this quietude rendering Diamang's past a "nonstory," on the contrary, this history became more interesting than ever. The absence of persistent, militant resistance revealed as much about the labor environment on the Angolan mines as its presence would have. In cases such as Diamang, when anticipated forms of worker resistance are faint, episodic, or, at times, even absent, professionalism facilitates an understanding of why African laborers engaged with colonial capital in ways that, at first glance, seemingly served *only* management's interests. Although historians of Africa have long been attracted to both sensational and more restrained examples of strife between labor and capital, the analytical utility of professionalism is most pronounced in scenarios—even in forced labor contexts—in which workers' objectives and employers' expectations overlapped in particular, meaningful ways. In these instances, the concept offers a more complex, and ultimately more accurate, understanding of labor relations than do presumptively antagonistic models or the reductive "collaboration-resistance" binary.

At Diamang, the constituent actions of workers' occupational and social professionalism inherently comported with company expectations, but shouldn't suggest a comprehensive submissiveness or lack of ingenuity. African employees creatively coped, and also periodically challenged company practices, flouted disagreeable policies, and articulated grievances. But, with the notable exception of desertion, workers did so while continuing to honor their occupational responsibilities and to interact in a socially conciliatory manner. Just as Diamang sought stability via a calculated, pragmatically inspired paternalism, from the 1930s onward most laborers and their families willingly, strategically played their complementary roles on the mines, in the company encampments and throughout the broader region. In this sense, corporate paternalism and worker professionalism were two sides of the same coin.

Third, due to the parsimonies of the Portuguese colonial project, Diamang could rely on coerced labor, or *shibalo*, to replenish its workforce long after these types of labor procurement methods were abandoned elsewhere in colonial Africa.[14] For almost five decades, residents of Lunda experienced forced labor in a way that was unparalleled in Portugal's African empire. Although the metropolitan government instituted shibalo throughout its colonies, nowhere else did it grant a concessionary company such exclusive access to labor. Diamang's early profits provided the motivation for the state, while, for regional residents, the installation of shibalo in 1921 immediately foreclosed avenues for alternative employment. Using the rich oral testimony I gathered from former workers and family members in Lunda, my interviews vividly illuminate the local forced-labor scheme and the ways it changed over time. In so doing, this book provides the first detailed account of the ways that Angolan men, women, and children experienced shibalo since the scathing, and ultimately internationally scandalous, 1925 Ross Report.[15]

Beyond providing a long overdue reexamination of forced labor in colonial Angola, the book also carves out new analytical space in this area. By the 1930s, Diamang's operational success and seemingly insatiable demand for labor meant that shibalo had effectively colonized Lunda residents' minds even before their formal engagement with the company, causing them to resign themselves to working on the mines for at least some period of time. Yet coerced recruits were also able to shape the local application of shibalo through a process of negotiation with Diamang that, on their end, required the same type of professionalism that they exhibited once on the mines. This meant forgoing the preemptive relocation strategies that they had embraced before the 1930s in favor of an insistence on reasonable treatment during their journeys to the mines. In this way, recruits contributed to an efficacious "culture of complaints" at Diamang, much as they would as full-fledged employees once they reached the mines. As such, within the first decade and a half of operations, recruits began *articulating* their objections, instead of using their feet to protest.

Fourth, the book also opens up new ways of thinking about women's labor roles, experiences, and strategies on colonial-era mines by highlighting their contributions to Diamang's phenomenal growth, as well as to the processes of social and familial reproduction that relocation to the mines threatened to interrupt. Although women never made up more than 5 percent of the remunerated labor force, they carried out tasks essential to the enterprise's day-to-day operations, including as cultivators on company plantations, as cooks and servers in mine kitchens, and as nurses and auxiliaries in company hospitals. Moreover, each evening these women performed an array of domestic tasks away from the mines in company encampments, including tending to any children present, retrieving water, and cooking. These professional and domestic responsibilities constituted a "double burden" that not only helped Diamang become the most powerful commercial entity in Portugal's African empire but also served to support and maintain these women's families in this new environment.

Heretofore, women have typically appeared in African mining historiography as beer brewers or prostitutes, or as cultural and/or social "buttresses" or repositories.[16] However, African women at Diamang largely operated outside of state- and company-defined realms of criminality, in part because urbanization was highly limited in Lunda. As such, they were not fringe actors involved in "devious" acts in the local informal economy, but rather vital members of the formal community of mine workers. In practice, what links geographically disparate communities of women on and around mines throughout Africa is their adeptness at exploiting the opportunities associated with their respective environments. At Diamang, women seized the opportunity to participate in remunerated labor and, just as their male counterparts did, over time became a highly productive, "professionalized" workforce. Indeed, despite—or perhaps because of—the myriad challenges women faced, they generally sought social and occupational stability by adopting a focused, committed approach, both before and after the whistle blew.

Fifth, the fellowship that marked Diamang's encampments challenges notions of these spaces as inherently violent and socially acrimonious.[17] Instead, I argue that the relatively uncrowded accommodations, Diamang's introduction of well-received, post-shift recreational activities, such as film screenings and athletics, workers' mutual tolerance—a central component of their social professionalism—and the stabilizing presence of women and children, collectively rendered the encampments sites of social accord. Unlike the discordant, all-male workforces that feature in Dunbar Moodie's and Robert J. Gordon's studies of mines in South Africa and Namibia, respectively, laborers at Diamang generally ignored, rather than regarded, ethnic, linguistic, marital, tenurial, or even age differences.[18] As José Coxi, a former employee, explained to me: "While . . . at Diamang, it was possible to learn others' languages. . . . It would happen while eating, drinking, etc. and we became better friends as a result."[19] In fact, even as encampments grew more demographically diverse, this miscellany produced more amities than enmities.

This book also builds on Moodie's and Gordon's work on social networks by suggesting that family members on Diamang's mines often assumed the supportive roles that all-male "brotherhoods" did elsewhere. Over time, the familial and extrafamilial cooperation that pervaded Diamang's encampments enabled inhabitants to enhance their lives through a range of strategic activities, including sharing domestic tasks, forging friendships, and both bartering and socializing with members of nearby communities who were otherwise unaffiliated with Diamang. Thus, both on and away from work sites, cooperation, or at least tolerance, trumped any antagonistic expressions of identity.

And, sixth, Diamang's virtually autonomous, extensive "state within a state" distinguishes it from mining operations elsewhere in colonial Africa. Although some colonial-era concessionary companies oversaw vast areas and others offered a range of services within smaller, more

confined spaces, Diamang's control was both expansive and comprehensive. Within Lunda, the company steadily shouldered the social
costs of production. In exchange for their labor, residents were over
time, for example, able to sell any agricultural surpluses to Diamang at
fair prices (a common refrain from informants); receive free health care
via the company's mobile health teams; and purchase low-cost provisions, including food, clothes, and sundry items, at company stores.
The diamond enterprise's hegemony throughout such a sizable area
and its unique tactics within this region are crucial to comprehending
how and why Diamang and its African labor force adopted and periodically adjusted their respective approaches to one another.

Diamang's commanding regional authority also deepens existing
understandings of the human and material resource inadequacies of
Portugal's broader colonial project. In practice, the company's operational area in Lunda bore little resemblance to Portuguese colonial
space elsewhere in Africa, and beyond. Within Diamang's expansive
concessionary zone, colonial administrators were extremely thin on the
ground and, by the 1930s, were essentially subservient to the industrial
behemoth, which, via a series of generous loans, had become one of
the state's largest creditors. Consequently, Diamang reserved for itself
many of the assistive aspects of regional governance, while leaving
the more disagreeable endeavors, such as coercive recruitment, to the
agents of the colonial state. Only a well-financed, pragmatic, and, ultimately, paternalistic entity—"a state within a state"—could, or even
would, have engineered this arrangement.

Diamang's "state" also had experiential implications for its white
employees. For example, the company offered its Portuguese staff, who
derived from a European "backwater," amenities that, for the most part,
would not have been accessible to them back home, including superb
health care, high-quality education, an array of recreational opportunities, and, in general, a standard of living that would have been hard to
duplicate in the metropole, or even elsewhere in Angola. Moreover,
these employees were not permitted to travel elsewhere in Angola during leave time, reinforcing the notion that Diamang's area was distinct from the rest of the colony. The company even went so far as
to establish a different time zone in its operational area to emphasize
its discreteness.[20]

This "state" in colonial-era Lunda also suggests that the way(s)
mining enclaves are understood be reconsidered. Although Diamang

established a geographic perimeter, it did not erect any physical impediments to delineate its territory, nor did fencing ring its mining encampments. Rather than considering every African employee a presumptive diamond thief, the way that colonial-era mining companies elsewhere approached their black—and, to a certain extent, white—employees, Diamang's approach differed. Although the enterprise's "borders" were certainly permeable, the company ensured that individuals abstained from illegal diamond buying and selling by relying on the threat of punishment, but also by carefully cultivating its relationship with regional residents. The success Diamang enjoyed as a result of its unique, more measured operational approach cannot be overstated, as it casts the scenario in Lunda in sharp contrast to that of Sierra Leone, Angola's colonial-era, alluvial diamond partner to the north. In that setting, the industry was plagued for decades by theft and illegal digging, buying, and selling—all of which presaged the eventual, internationally notorious violence. Conversely, in Angola, a "wild west" environment only began to materialize on the edges of Diamang's tightly controlled "state within a state" in the waning days of its operations.

In addition to opening up new ways of thinking about African mining contexts, this book also acts as an historiographical foil to the substantial body of scholarship that examines mining in South Africa. Absent from the Angolan mines were the factional, often ethnically based, "politics of difference" that played such infamous roles in shaping South African mine workers' experiences, as well as the cramped, male-only compounds that Wilmot G. James, J. K. McNamara, and others have suggested helped to provoke this violence.[21] Similarly, scholars' characterizations of labor relations on South African mines as "lacking compassion, antagonistic . . . and wasteful," which, in turn, prompted the creation of powerful trade unions, stands in sharp contrast to Diamang's investment in its African workforce.[22]

The Angolan company's pragmatic paternalism and the complementary professionalism that its African workers embraced largely explain this experiential divergence. Certain, context-specific realities, though, are also analytically relevant. At Diamang, for example, ethnic homogeneity (roughly three-quarters of its African laborers were ethnic Chokwe), alluvial deposits (versus the much more dangerous, underground mining in South Africa), and a regime in Lisbon that prohibited trade unions of any sort—black or white—partially explain the absence of social and labor unrest on the Angolan mines. Yet Diamang and its African labor force are ultimately responsible for generating the regional quietude. For example, although the Angolan enterprise periodically considered adopting the South African compound system and regularly employed mine managers who had prior experience in South Africa, it repeatedly rejected that housing model in favor of relatively spacious accommodations. I contend that this decision was a major contributing factor to why Diamang, unlike its South African counterparts, constitutes an example of an institution that succeeded in cultivating a set of labor relations that enhanced productivity, production, and, ultimately, profits.

METHODOLOGY AND SOURCES

This book is one of only a handful of recent historical monographs that focus on Angola, and thus it contributes to the nascent process of drawing aside the curtain of the country's past.[23] In great part, this scholarly absence is attributable to the roughly four decades of conflict that ravaged Angola from 1961 to 2002, beginning with the War for Independence (1961–75) and followed immediately by the protracted Civil War (1975–2002).[24]

Consistent with the dearth of recent studies on Angola's history is the virtual absence of scholarship on Diamang.[25] Despite its importance to Angola's past, and present, this book is the first to reconstruct the histories of Diamang and its African labor force. Following the civil conflict, the Angolan government restricted access to Lunda, which remains today a violently contested area. This combination of prolonged conflict and ongoing violence arrested the historiographical process and virtually silenced a whole generation of Angolans. The interviews I conducted after finally gaining permission to carry out research in the area render it the first to capture the voices of the African men, women, and children who once labored for Diamang.

This study also draws on archival evidence to reconstruct the lives of these employees and family members who traveled to, lived, and labored on Diamang's mines. Colonial records accessed in Portugal and Angola include reports and letters issued from colonial administrators in the field and labor contracts concluded between the state, the company, and individual workers. However, the extant Diamang records, now deposited at the Universidade de Coimbra, in central Portugal, constitute the most important written materials.[26] This invaluable collection of company documents provides both quantitative and qualitative information related to workers' social origins, recruitment, housing, diet, ailments, and the labor process. In particular, unpublished internal company documents, including letters, telegraphs, memos, and mine inspection reports, illuminate the challenges that Diamang faced as it continually reorganized its expanding African workforce, as well as the strategies that both the company and the labor force employed to overcome them.

During the course of my research, I also collected and digitized roughly a thousand black-and-white photographs of daily life at Diamang, some of which are incorporated in the text. These images provide a rare glimpse of colonial Angola and, in particular, Diamang's mines, as the company strictly forbade private photography in its operational zone. I encountered many of these photographs in unpublished company reports and documents, but many others were produced by informants who had absconded with them following the company's abandonment of its offices in Dundo in the face of the encroaching civil conflict. Remarkably, many of these photographs were stored in old shoeboxes in informants' houses but remain(ed) in excellent condition.

Despite the richness of these visual and archival materials, they are limited in two significant ways. First, they almost never penetrate life after the whistle blew or during recruits' journeys to the mines. This limitation hinders a reconstruction of social and gender relations, as well as extra-work-site activities, which, in turn, hamper an investigation into the strategies that Africans pursued in spaces beyond the company's gaze. Second, many of the enterprise's documents have been lost or destroyed, and, thus, the collection of Diamang records housed at Coimbra is incomplete. The absence of these materials grants the archive an uneven quality and suggests silences in the written record that may have, in fact, existed, but may also have resulted from this deficit.

Because of the constraints of the written record and because this study considers the lived experiences of male and female laborers, oral testimony figures prominently in the narrative. The oral evidence on which this study relies includes more than eighty one-on-one interviews of former Diamang workers and family members conducted almost entirely in Lunda over the course of approximately three years (2004–6). These interviews yielded invaluable interior accounts of daily life on the mines, filled many of the holes in the written record, and often belied published company reports that, intended for shareholders' eyes only, minimized the challenges Diamang faced on the ground.

Collecting these testimonies presented serious logistical and security issues. Angola's Lunda Norte Province, which encompasses virtually all of Diamang's former operational area, is a highly secure, secretive, and violent region due to the vast wealth its diamonds continue to generate. To gain access to the area to interview former company employees I had to establish a series of key relationships with senior government officials, clergy, and private mining company personnel.[27] This process lasted months and even then never guaranteed that local officials would allow me to travel to Lunda Norte or deplane upon arrival. Other times, I was detained after arriving, and my passport subsequently disappeared into the murky layers of Angolan officialdom only to reappear weeks later.

Underpinning every one of these episodes was the culture of fear, suspicion, and paranoia that pervades contemporary Angola, and especially the violent, "wild west" stretches of Lunda Norte Province, in which well-armed groups vie for control of the area's lucrative diamond-bearing soils. Thus, beyond logistical and security concerns, often my most daunting task was to convince the Angolan officials who obstructed my work of the importance of recording the experiences of the dwindling number of colonial-era mine workers still living and, more broadly, of the value of this study in an area where no historical research had been carried out for decades, if ever.

Collecting oral testimony in this context also presented difficulties of interpretation. The conditions at Diamang that former workers described reflect the current social realities on the ground, requiring that their personal narratives be read just as critically as colonial documents. In fact, many informants delivered testimony that their actions at Diamang problematize or even contradict. For example, in post-conflict Angola, any discussion of forms or expressions of violence—past or present—by the war-weary population is (understandably) unpleasant, and, thus, former mine overseers often elided the physical abuse that they committed against fellow Africans. Virtually all informants did, however, express a genuine nostalgia for the "time of Diamang," a period marked by tranquillity and certainty, though the post-independence chaos that has plagued Lunda undoubtedly heightened these feelings.

Workers also offered testimony that emphasized victimization by the Angolan diamond parastatal, Endiama, following independence. As many of my informants were actively engaged in a legal struggle with Endiama over back wages, it was politically expedient for them to stress this exploitation. Concomitantly, they highlighted the relatively agreeable working conditions at Diamang in an attempt to contrast the postcolonial enterprise with its predecessor. Informants engaged in this politics of memory speculated that because of my interest in their histories I might be able to advance their current cause. Set within this contemporary context, these oral documents reflect informants' discontent with their ongoing marginalization and exploitation, which they deem the post-independent Angolan state is supposed to safeguard against.[28] For these individuals, my historical project highlighted just how much was currently at stake.

ORGANIZATION: TO THE MINES AND BACK

Following this introductory chapter and an ensuing chapter that examines the political economy of colonial Angola, the book follows African recruits and any accompanying family members from their villages to the mines and then back again. Drawing on the archival and oral evidence outlined above, I highlight change over time within each episodic chapter, so as to provide a diachronic understanding of the environments on and around the mines and the shifting ways that workers interacted with the company, and vice-versa.

Chapter 2 provides essential information on the political economy of colonial Angola and the history of the Lunda region both before and during the company's incipiency, as well as a detailed account of the evolving state-company relationship in key areas such as labor recruitment, revenue sharing, and security. Using this foundational information, readers can better situate the company and its African labor force in the shifting theater in which they operated, while the chapter also provides context for the proceeding thematic chapters.

Chapter 3 examines Diamang's relentless efforts to secure both volunteer (*voluntário*) and forced (*contratado*) laborers and these individuals' divergent recruitment experiences. Further differentiating this process for coerced recruits was their distance from the mines, while recruits who had wives and/or children were also forced to decide if these family members would accompany them or not. The key oral and archival evidence presented implicates a series of actors in the durable coercive recruitment scheme in place in Lunda, including local Portuguese colonial administrators, sobas (traditional authorities), and African colonial police (*cipaios*). To explore both the entities involved and the variance in recruits' experiences, this chapter makes extensive use of oral testimony to trace individuals' trips from their villages, to company headquarters at Dundo for processing, and eventually to the mines. The chapter also examines recruits' growing praxis of professionalism, evinced in the range of strategies they employed over time to improve the overall recruitment experience.

Chapter 4 extends the examination of recruits' daily experiences by following them to their respective work sites. In particular, the chapter explores the changing labor processes in which these workers engaged and focuses on the fitful and uneven impact of the company's gradual efforts to mechanize portions of its operations; the shifting size, composition and organization of the workforce; the different tasks that Africans performed; daily work schedules; the manner in which African and European overseers supervised the labor force; the occurrence and aftermath of occupational accidents; and worker compensation. The chapter also examines women's and children's labor experiences and the vital contributions that they made.

Chapter 5 explores the ongoing negotiation process between the African workforce and Diamang through the prism of their respective labor strategies. Both male and female laborers mitigated the daily work regimen by, for example, sharing tasks, singing songs, starting work early to minimize exposure to the afternoon sun and leveling complaints about particularly objectionable labor practices or individual supervisors. Some male employees even sought to either temporarily or permanently avoid the daily regime through a range of tactics, including feigned illness, absenteeism, and desertion. However, Diamang's increasingly effective paternalism—including company officials' escalated responsiveness to workers' grievances—and laborers' corresponding levels of occupational professionalism saw these riskier strategic pursuits decrease dramatically over time.

Chapter 6 draws heavily on oral testimony to explore workers' post-shift lives, thereby providing an interior view of their experiences after the whistle blew. Laborers' challenges did not end with the shift, as commutes to and from work sites each day could reach dozens of kilometers, and when they arrived home, water had to be retrieved and dinner prepared. Over time, however, the company's housing upgrades, achievement of food security, introduction of leisure activities, improved health care services, and broad tolerance of workers' evening endeavors helped temper the demanding nature of life on the mines. It was in this increasingly agreeable context that workers and family members pursued a range of strategies intended to further improve their post-shift lives, including, most significantly, entering into an array of social relationships that helped foster a stable and amiable environment in company mine encampments.

Chapter 7 examines the culmination of coerced laborers' engagements with Diamang, from the expiration of their contracts and their decision to remain on the mines or return home, to the ways that working for the company shaped their post-Diamang lives. Employees who opted to return to more familiar settings, as most did, commenced the process of "repatriation" (as Diamang referred to the trip home). Informants described the concluding moment of this process fondly, reuniting with friends and relatives and almost invariably reintegrating seamlessly into village life. Subsequently, many workers remained home, never returning to the mines, while others reengaged—either willingly or unwillingly—several times.

Finally, an epilogue traces the events that transpired in Lunda following Angolan independence; examines how these developments affected Diamang, its former employees, and the newly independent nation; and limns the contemporary implications of this history, thereby bridging the past and present. It also outlines what's at stake by revisiting these increasingly distant decades of mining in Lunda. The same deposits that Diamang's African laborers had once excavated continued to play a major role in postcolonial Angola, eventually generating so-called "blood" or "conflict" diamonds. Although Diamang ceased meaningful operations shortly after 1975, the mineral wealth buried in the Lunda soil helped fund the civil war that acutely destabilized the newly independent country. With the spread of the conflict into Lunda, the cacophony of gunfire and mortar rounds violently concluded the stability that Diamang and its African labor force had for so long maintained. Ultimately, these deadly sounds reverberated far beyond Lunda's borders, infamously catapulting the region and its "blood diamonds" onto the global stage.[29]

2 ᔰ A Bountiful Place
The Political Economy of Lunda, 1870–1975

It can be said that the eradication of our colonial deficit depends on the concentration of our efforts to exploit extensively the natural riches of our colonies, namely Angola.

— *Portuguese Colonial Council, 1912*

The black population of Lunda neither worries nor inspires us [to act]. We have at work almost 25,000 natives and are convinced that they prefer to live with us in peace and good harmony.

— *Diamang director Ernesto de Vilhena, following the outbreak of revolutionary violence in Angola in 1961*

ALTHOUGH LUNDA EVENTUALLY BECAME "a state within the state," the Portuguese colonial state was instrumental to Diamang's success. Even before the company's arrival, the colonial army was laying the foundations for Diamang's operations through a series of pacification campaigns in Lunda.[1] Determined and well-armed local groups fiercely resisted these offensives, but by the mid-1920s the Portuguese military had largely subdued any remaining rebellious communities. This military success provided the operational security the company desired and effectively transformed Lunda into the secure labor reservoir that Diamang so desperately needed. Although the comments made by Ernesto de Vilhena in the epigraph to this chapter mask some intracompany anxieties in the post-1961 period, the regional population did, indeed, appear to "prefer peace and good harmony."

From the 1920s onward, a succession of unique agreements between the company and the state that covered revenue sharing, taxes, security, and access to labor further improved the already favorable operating conditions that the colonial political economy offered and enabled Diamang to operate with virtually no statutory oversight. As Joaquim António Issuamo, who worked at Diamang from 1959 to 1975, explained: "From the workers' perspective . . . the government did not care about us. . . . The government of the worker was Diamang."[2] Over time, Diamang's mounting profits dramatically enriched the colonial state via a series of profit-sharing agreements and loans, cementing the mutually, fiscally beneficial relationship.

Despite the generally amicable relations between the company and the state, divergent objectives also generated moments of tension. For example, the company's substantial labor demands uprooted Lunda residents, undermining the fixed tax base that the state strived so hard to develop in this isolated region of Angola. Local colonial administrators also occasionally strained state-company relations by pursuing personal agendas that hindered both recruitment and the recapture of workers who had deserted.

In this chapter, I provide an account of the Lunda region immediately before the start of mining operations to highlight both local resistance to colonial encroachment and the state's contributions to Diamang's eventual economic success. After outlining the political economy of colonial Angola, I trace the evolving state-company relations in the areas of labor, revenues, and security from the enterprise's inception through Angolan independence. I contend that the exceptional agreements around which this relationship revolved shaped the histories of both the company and its African labor force much more profoundly than did the broader political economy in colonial Angola. I also argue that the late 1930s marked the most transformational period in the state-company relations, as, after this time, substantial profits enabled Diamang to assume an equal, if not superior, position in its dealings with the colonial regime. This wealth deepened the relationship between the state and its new benefactor and paved the way for Diamang to pursue its pragmatic, profit-fueled paternalism in Lunda, which stood in sharp contrast to the underfunded colonial state's penurious approach to the indigenous population elsewhere in the colony. Finally, I provide an account of the period immediately following the commencement of mining operations, which highlights local residents' reluctance to engage with Diamang, as well as the initial logistical and managerial challenges the company faced. This examination is also intended to serve as a foundational piece on which the ensuing chapters will build.

LUNDA ON THE EVE OF THE DIAMOND RUSH

Decades before the arrival of Diamang and the colonial state in Lunda, the region had been a flourishing commercial center. Local Chokwe populations had *not* been actively mining diamonds, on which they placed little to no value, but rather were engaged in the commerce in slaves, ivory, wax, and, just before Diamang's arrival, rubber.[3] These Chokwe traded with other African communities, as well as with *sertanejos*—the African or *mestiço* (mixed-race) agents of Portuguese merchants—and later, with these merchants directly.[4] In addition to trading locally produced items, Chokwe communities just as often acted as intermediaries in long(er)-distance trade, applying levies on commercial caravans passing through the area. These extensive revenue-generating activities enabled local populations to acquire copious firearms and also instilled in them a sense of independence that presented considerable challenges to Diamang as it struggled to launch and expand its operations. Even with the rubber trade in precipitous decline shortly before Diamang's arrival, colonial troops and company officials encountered a well-armed population bolstered by a history of commercial success rather than a destitute and defenseless population. This remote region of Angola was consequently the last to be "effectively occupied" by the Portuguese.

PRE-DIAMANG LUNDA: DIAMOND-LESS WEALTH AND PRIVATION

If one word could be used to summarize life in Lunda immediately before the start of mining operations, it would be *rubber*. As far back as the 1870s, African merchants from Angola's central highlands were purchasing rubber from the Chokwe, and by 1887 this commodity had become Angola's leading export.[5] This commercial activity, in turn, enabled the Chokwe to secure the firearms they needed to rout the previously dominant Lunda peoples and establish themselves as the foremost group in the region. Profits to be made from rubber also temporarily minimized population outflows from Lunda, despite the growing need for laborers in neighboring Northern Rhodesia (Zambia).[6]

Angolan rubber exports peaked over the course of the next decade before prices began to decline, slowly at first, and then abruptly in the face of Malaysian plantation rubber. From 1910 to 1917 alone, prices fell by approximately two-thirds, while communities were simultaneously exhausting local supplies.[7] Although regional residents tried to compensate by increasing wax and hide exports, as well as agricultural goods, including corn, beans, manioc products, pumpkins, and potatoes, these items generated only paltry revenues compared to those streaming in during the height of the rubber boom. A 1916 colonial official's report from the field declared, "Regarding the inability of exports to improve local populations' conditions, rubber, the principal good exchanged to that end, has remained almost valueless. Hides and wax have continued to maintain good prices but the quantity of these items exchanged is small."[8]

Prior to the decline in rubber production and profits, Chokwe residents had been able to procure foodstuffs, as well as guns and gunpowder. They originally purchased this materiel from either itinerant Portuguese merchants or, later, from merchants who had established trading houses in the western stretches of Lunda. Although many of the firearms they acquired were of dubious quality, or even antiquated, the quantity was considerable. According to one estimate, from 1909 to 1912, African populations in Angola had purchased over 115,000 American Civil War–era rifles.[9] After the colonial government banned the sale of these items to Africans in 1913, Chokwe traders simply traveled across the porous Belgian Congo border and purchased them at trading posts deliberately erected just over the colonial boundary. In addition to losing valuable rubber supplies to these neighboring merchants, Portuguese colonial officials also rued the further arming of local populations. A 1917 colonial official's report from Camaxilo, in western Lunda, captures this chagrin and the difficulties that this arrangement was causing. "In this way . . . the majority of the rubber made in our territory is not sold within it. . . . Though the sale of gunpowder is prohibited in Angola, the heathens' acquisition and possession of as much as they desire cannot be prevented; and beyond this evil, there is another—the diminishment, or even death, of our commerce up to the Cuango [River]. . . . And beyond that evil there remains a moral evil, as the heathens take the fact that we will not sell them gunpowder as a sign of weakness."[10] The ongoing procurement of guns and gunpowder by regional Chokwe communities had serious implications for colonial troops trying to pacify the local population and Diamang officials trying to set up mining operations in the wake of the Portuguese military advance.

THE ESTABLISHMENT OF DIAMANG AND THE COMMENCEMENT OF FORMAL MINING OPERATIONS

The Companhia de Pesquisas Mineiras de Angola (PEMA), which began limited prospecting in 1913, was responsible for the first measures aimed at gaining access to the riches buried in Lunda's soils and riverbeds. After prospectors from Forminière, a concessionary diamond company operating just over the border in the Belgian Congo, had verified the presence of diamonds in Lunda in 1912, a multinational group of investors formed PEMA on September 4, 1912, to realize the anticipated bounty.[11] Despite the formidable, and even lethal, challenges that PEMA prospectors encountered, over the next few years they managed to identify a series of diamond deposits along the Chikapa, Chihumbe, and Luembe rivers with the potential to yield over three million carats of high-quality stones.[12] Realizing that significant capital and resources would be required to fully exploit these deposits, PEMA officials astutely comprehended that a better-funded and more focused enterprise would be necessary.

To satisfy this objective, a diverse and powerful group of international investors formed Diamang on October 16, 1917. This team of financial backers featured many of the original PEMA investors, including two from Portugal, the Banco Nacional Ultramarino (BNU) and Henry Burnay & Co., while new investors included The Société Générale de Belgique; the Mutualité Coloniale; La Banque de l'Union Parisienne (Paris); O Sindicato de Diamantes de Londres; the Anglo-American Corporation of South Africa; the Morgan Bank; the Oppenheimer Group; and the American financiers Ryan and Guggenheim. According to Diamang's corporate configuration, Belgian, English, French, South African, and American entities commanded 80 percent of the shares, and the Portuguese state the remaining 20 percent.[13] To strengthen the Portuguese connection, the investment team designated Lisbon as the site of the company's headquarters, with supporting offices to be established in London and Brussels. By June 1918, PEMA had transferred all of its rights to diamond deposits in Angola to Diamang, thereby finalizing the new company's monopoly.[14] Diamang commenced operations with a respectable 90,000$00 escudos

(£20,000) in capital, but by the end of the 1920s, based on the company's initial successes and the fact that investors had, early on, begun to see positive returns on their money, this figure ballooned to 9,000,000$00 (£2,000,000).[15]

OPEN FOR BUSINESS: THE POLITICAL ECONOMY OF ANGOLA, 1917–75

Diamang's operational era spanned two metropolitan regimes, which varied only in the degrees to which they endeavored to foster highly favorable commercial environments for enterprises operating in the colonies. Given Portugal's dismal fiscal situation throughout the colonial period, it eyed its African territories as potential sources of revenue, and none more so than Angola. Consequently, both the idealistic Republican government (1910–26) and the ruthlessly sparing Estado Novo regime (1926–74) worked to create political economies in which the pursuit of profits justified even the most egregious treatment of indigenous laborers.

In its roughly sixteen years in power (1910–26), Portugal's Republic could point to a series of modernizing, "commercial friendly," and highly expensive achievements in Angola. These accomplishments included the replacement of military officials with civil servants in local administrative outposts (*postos*), the successful completion of pacification campaigns (including in Diamang's operational area), and the dramatic expansion of the road network.[16] Norton de Matos, who served as Angola's governor general from 1912 to 1915 and then as occupant of the newly minted post of high commissioner from 1921 to 1923, was undoubtedly the most zealous and liberal Republican pursuant of these often well-meaning but fiscally irresponsible reforms. Indeed, most of his successful initiatives were facilitated by substantial foreign loans rather than financial support from the Portuguese state. As Pitcher argues, "The Republicans preached a closer relationship between the colonies and the metropole, yet failed to develop or finance it. They promised state support yet extolled the virtues of decentralization. . . . They outlawed slavery, but introduced forced peasant production."[17] When the colony was unable to meet the loan repayments associated with Norton de Matos's projects, the metropolitan government recalled the foresighted, yet ultimately imprudent, administrator and leveled malfeasance charges against him.[18] In the wake of the mounting fiscal disaster in Portugal and the attendant public discontent, a military coup d'état in the metropole in May 1926 removed the Republican government and ushered in Dr. António Salazar's durable, dictatorial Estado Novo (New State).

The new regime was even more business-friendly than the preceding administration. Almost immediately, it strove to cultivate an inviting commercial climate in Angola by facilitating access to cheap, bound African labor, granting concessions to both Portuguese farmers and industrial interests, and expanding the colony's infrastructure.[19] In response, waves of Portuguese settlers ventured to Angola to take advantage of the new social arrangement in which most of black Angolans officially became "*indígenas*" ("natives"), a legally marginalized and violently exploitable subsection of colonial society.[20] The New State also ended the fiscal and political autonomy that administrators like Norton de Matos had enjoyed by centralizing administration of the colony. Lisbon, rather than Luanda, is more accurately understood to have been Angola's administrative capital throughout the era of the New State.

With rural Angolans increasingly available to meet the labor demands of colonial capital, several profitable agricultural and industrial enterprises emerged, including, most notably, diamonds, and by the 1950s, coffee.[21] Even after Angolan revolutionaries began to fight for independence in the 1960s, propelled on by the New State's oppressive policies and the "winds of change" sweeping across the continent, the colonial economy churned on, protected in part by the Portuguese military and an array of police organizations. In fact, the Angolan economy grew dramatically during the 1960s, suggesting that while the independence movements may have struck fear in the hearts of many white settlers, they had virtually no impact on an economy that was beholden to the uninterrupted flows of African laborers to and from work each day.

A series of inspections conducted in the 1960s and 1970s by the International Labour Organisation (ILO) was marginally more effective in shaping the labor regime in the colony.[22] Primarily through duplicity and indifference, the Portuguese regime had weathered a series of incriminating reports earlier in the century about labor conditions in its colonies. Although these critical accounts engendered a flurry of international condemnations, they had done little or nothing to improve conditions for African workers on the ground.[23] Given the altered continental and international contexts, however, the ILO investigators' reports proved to be more difficult to dismiss.[24]

Over the course of the ILO's examinations of labor conditions in Portugal's African colonies, representatives visited Dundo and other locations within Diamang's concessionary area and ultimately called both company and colonial officials to testify at an international tribunal related to the controversy. Diamang officials' unwavering confidence was regularly on display during the judicial process. In 1967, for example, Guilherme Moreira, Diamang's delegate administrator, countered accusations that workers whose provenances were further removed from the company's mines disproportionately performed the more difficult mining tasks by likening them to recently arrived immigrants in the United States or Europe, who also typically work(ed) more difficult jobs.[25] Yet although Diamang officials could facilely refute allegations that, for example, "in 1954, the diamond mines in . . . Lunda employed *eleven million* workers, a quarter of which were forced laborers," the investigations and subsequent proceedings did prompt the colonial state to remove local administrators (*chefes do posto*) from the coercive shibalo labor scheme, thereby partially dismantling it.[26] However, most state-sponsored reforms were simply cosmetic, and, despite the unwanted attention that the ILO inquiries had generated, the company's paternalistic relationship with its African workforce was undisturbed, permitting the enterprise to continue to rely on indigenous labor to mine for the riches buried in the Lunda soil.

A DURABLE PARTNERSHIP: DIAMANG AND THE COLONIAL STATE

Shortly after Diamang's formation, the company and the Portuguese colonial state consummated several agreements, most notably in 1921. These exceptional concords secured a portion of Diamang's revenues for the state and, in return, allotted the company an expansive concessionary area, exclusive access to indigenous labor, tax-exempt status, and the state's assistance in both regional security and the suppression of the illegal commerce in diamonds. Over time, these agreements (and the prodigious diamond revenues on which they were predicated) afforded Diamang benefits well beyond those that the already commercially friendly climate offered. The pacts also insulated the enterprise against "rogue" colonial officials' periodic queries regarding the company's operations and provided Diamang with legal cover in disputes with local colonial administrators during the early years of mining operations. In practice, the specific areas of cooperation covered in these agreements—labor, revenues and security—played a larger role over time than did the broader political economy in the colony in shaping the lives of the company's African employees and family members.

From the inception of Diamang's mining operations, the company relied on state officials to help it procure African laborers. This collaborative endeavor was largely ad hoc in nature until company officials and Norton de Matos inked the far-reaching 1921 agreement (Decree 11). The colony's fiscal precariousness propelled Norton de Matos to sign, while shortages of the manpower required to expand operations constituted the company's primary impetus. Prior to the agreement, senior company officials had been complaining to Norton de Matos that local administrators were regularly reneging on promises to furnish laborers and that the consequent shortages were hindering production. For example, Antonio Brandão de Mello, a Diamang official, wrote Norton de Matos in April 1921, lamenting that "[last year], the Company was only able to get one-quarter of the manpower it needed for its operations, and we foresee the same for the current year. The Governor of Lunda, Oliveira Santos . . . informed me that he would round up forced laborers and that . . . 1,500 would be distributed to the Company, yet none have been furnished thus far."[27] Shortly thereafter, as part of Decree 11, Norton de Matos designated the entire district of Lunda an exclusive recruiting zone for the company, instructing chefes do posto to "intervene directly in the procurement of contratado recruits and to send these recruits to Diamang—and *only* to Diamang."[28]

Following the directive, a roughly twenty-year period ensued in which chefes do posto dutifully rounded up contratados for the company, while openly complaining about the tax revenues they would lose if these laborers relocated to the mines permanently—exactly the scenario the company was trying to engineer. Periodic rows erupted when individual chefes, who were occasionally supported by more senior colonial officials, failed to cooperate fully. For example, a 1929 letter from the governor of the District of Lunda informed Diamang officials that "my functionaries (*chefes do posto*) have neither the time nor the will to help the Company recruit, even though I acknowledge that Decree 11, consummated by the Company and Norton de Matos, requires it."[29] Chefes do posto could also be begrudging participants in the obligatory allocation of their cipaios (African policemen) to engage in the often fruitless process of trying to capture recruits or workers who had deserted. In response, the company issued a series of missives. A letter written by James R. Evans, a Diamang engineer, in 1930 bemoans the fact that "the worst feature about contratados deserting is the indifference of the proper authorities towards this matter after we inform them of these desertions. Their authority seems to be waning or they do not care to use it after they have collected the *imposto* [annual tax]."[30] On other occasions, company officials complained about deficiencies in the overall administration of the colony, rather than individual chefes. In 1934, Diamang representative Jorge Figueiredo de Barros sent a letter to the governor of Malange (a region immediately west of Lunda) admonishing the administration for the problems that periodically vacant postos caused. "The number of voluntários . . . has diminished as of late because of the lack of pressure to pay the annual tax, due solely to the unmanned administrative postos, and specifically, those at Sombo, Luia, Canzar, Lóvua, and, soon, Cachimo."[31]

Whether or not accusations like Barros's were accurate, by the end of the 1930s the company's elevated sway over senior colonial officials obviated the need to level them. As Diamang's profits soared, it was increasingly able to apply pressure on senior colonial officials in Luanda and Lisbon—including even the Portuguese dictator, Salazar. For example, in 1963, when the governor general of Angola proposed to end the company's monopoly on labor in Lunda, Diamang's director, Ernesto de Vilhena, appealed directly to Salazar. "I am beginning to grow tired, not physically, but intellectually, as a result of having to deal with so many stupid and disloyal people."[32]

In response to Diamang's mounting financial importance to the state, the metropolitan government mandated compliance in the field and dismissed chefes who failed to cooperate. Joaquim Trinidade, who was with Diamang from 1965 through Angolan independence, described the nature of this relationship: "There were good relations between the state and company officials. The company gave local colonial officials end-of-the-year baskets and furnished houses! . . . [Yet] if a chefe de posto abused his power, the company could have him fired or moved outside of its concessionary area."[33] Over the decades, Diamang and the state officially reaffirmed the landmark 1921 agreement several times, but without the overt discontent emanating from the corps of local administrators that had followed the initial decree.

The company's successful browbeating of these local officials ensured that its vital regional labor pool was effectively safeguarded.

REVENUES: DIAMANG AS PROVIDER

As a result of the state's willingness to supply workers to Diamang, it derived several financial benefits from the company that proved to be crucial for the fiscally floundering colony. In fact, the emanation of any revenues from formerly moribund Lunda was a truly welcomed development. Just before Diamang's arrival, the Lunda governor had lamented that "the insignificance of the receipts that the District of Lunda collects, versus the expenses that it generates, makes it weigh heavily on the financial imbalance of the Province. In the rudimentary state in which it finds itself, the district lives almost exclusively from the subsidies that Luanda sends it, with . . . a coffer normally empty and a great number of creditors to bother it. The acute crises that the district has endured due to a lack of funds have, unfortunately, not been rare."[34]

The ensuing fiscal windfall for the state came in the form of company shares, a portion of Diamang's profits and company-issued loans.[35] The sweeping 1921 agreement called for the diamond enterprise to grant to the colony 40 percent of its net profits and 100,000 shares of company stock and extended Diamang's concession, which had originally been set to expire in 1923, another twenty-eight years, to 1951.[36] Moreover, the pact formally exempted Diamang from all current and future tax obligations on profits, imports, and all of its (diamond) exports — an arrangement not even enjoyed by missionary organizations operating in the colony.[37] As part of this mutually beneficial agreement, Diamang also increased its capital investment so as to "more extensively exploit already recognized reserves and to locate new ones." Collectively, these contractual stipulations translated into immediate and immense financial dividends for the state; by one estimate the company paid into the colony's treasury some £1,233,247, just between 1921 and 1923.[38]

With the state and company both enjoying this arrangement, in May 1937 the former reasserted its commitment to Diamang by extending the diamond enterprise's exclusive concession from 1951 to 1971. In exchange, Diamang agreed to raise the portion of its profits that it shared with the colonial state to an even 50 percent. In May 1946, Diamang also opened a 100,000,000$00 escudo line of credit for the state, repayable at a modest interest rate of only 1 percent. A June 1963 agreement extended this amount to 150,000,000$00 escudos, repayable over twenty years.[39]

Diamang's escalating revenues made all of this financial largesse possible. Over the first dozen years of the enterprise's operations (1917–29), the total number of carats sold rose from approximately 4,100 to 312,000. According to the company, "During this period, the amount delivered to the Government of Angola, in compliance with the contractual dispositions, including dividends, portions of profits and loans, totaled £989,027,711. . . . No other enterprise installed in the Portuguese Empire . . . can compare."[40] Going forward, contributions to the colonial state's coffers continued to escalate, reaching an estimated £14.5 million by 1955 and surpassing £50 million by 1975. By that time, Angola had become the fourth largest diamond producer in the world.

SECURITY: A MUTUAL EFFORT

While the contractual relationship between Diamang and the colonial state roughly constituted an exchange of profits for laborers, both entities took an active—if less binding—role in the provision of various forms of security in Lunda. Initially, "security" consisted solely of the Portuguese military's efforts to subjugate the local population. However, during the next moment of acute security concerns, the period following the outbreak of the Angolan revolution in 1961, both Diamang and the state mobilized resources to defend the concessionary area. In fact, because both parties benefited from steady and seamless cooperation in this area, security-related endeavors produced relatively little friction between Diamang and the colonial regime.

In Lunda's "frontier environment," security was paramount to Diamang as it commenced commercial operations, yet spirited local resistance initially threatened the viability of its intentions and hindered early efforts. Of particular concern to both the state and the company was a "rebellious" soba named Calendende. Colonial officials estimated that, via rubber sales, Calendende and his followers had amassed thousands of rifles and over thirty-five tons of gunpowder, thereby rendering their ability to repel Portuguese advances and humiliate the would-be colonizers seemingly irrepressible.[41] Two of his more audacious exploits were the destruction of a colonial posto in 1908, after which he allegedly used the cranium of a dead Portuguese official as a vessel from which to drink a preferred local intoxicant, and the 1917 "massacre" of twenty

African soldiers attempting to recruit workers. Calendende enjoyed significant local prestige well into the 1920s, as he and his followers continued to violently impede Portugal's designs on the territory. In fact, even after Calendende's death, Diamang officials could be heard complaining to the governor of Malange that "the area between Cuilo and Lovua is [still] populated by soba Calendende's people, who always objected to the work [regime] and, until a few years ago, were still hostile to Europeans. . . . The area remains intoxicated by rebelliousness."[42]

To help effect the long-term regional stability that Diamang craved, the company often assisted the colonial government with its military undertakings. Although Diamang did not initially possess its own security forces, it could still provide valuable support. On at least two occasions in the 1920s, for example, company officials supplied the colonial state with useful information concerning indigenous populations and local geography ahead of Portuguese military advances.[43] Other times, Diamang officials appealed for targeted strikes against individual sobas who were hampering operations and/or attacking white employees. In June 1921, Brandão de Mello, a company official, wrote to Portuguese army commander Col. António J. Santa Clara, beseeching him to "take action against the soba Cunza, as he has risen in the past and causes paralyzation on the mines when he rises up. . . . Beyond that, the Company is genuinely concerned for the safety of its white employees."[44]

While Portuguese military operations were vital to both Diamang's short- and long-term goals, the suppression of local resistance also occasionally disrupted mining operations and drove away potential recruits, thereby generating local state-company acrimony. Just before penning the aforementioned letter to Colonel Santa Clara, Brandão de Mello had informed Norton de Matos that "the military operations undertaken in Lunda this past year have not brought about an improvement in the situation. To the contrary . . . military forces drove away laborers that worked in this area, halting our operations there for some time."[45] During the previous year, Diamang's managing engineer, G. H. Newport, had voiced a similar grievance to the governor of Lunda, explaining that "throughout 1920, more workers were available than ever before and the work had been running smoothly until the commencement of military operations."[46] Following these maneuvers, local sobas had prevented Diamang from constructing roads, realizing that these thoroughfares would facilitate the movement of troops and, eventually, the carving up and prying open of their formerly sovereign territory.

Over time, Diamang officials and Portuguese commanders overcame their differences by agreeing to schedule military operations largely according to company requests. This coordination made potential disruptions more predictable and reduced the overall numbers of workers lost. For example, because laborers were more readily available during certain stretches of the local agricultural cycle, namely January to July, company officials generally exhorted the army to carry out its offensives outside of this period so as not to scare away potential recruits. Other times, Diamang simply requested that the military conduct its maneuvers removed from the company's operations. According to Lute Parkinson, an American prospector in Diamang's employ: "One of the early prospectors, when working in the vicinity of Cavuco not far from the Congo border, was bothered when military operations between the Portuguese authorities and the local Chokwe suddenly broke out not far from the scene of his operation. He wrote a note to the sergeant in charge of the troops strongly suggesting that the war be moved away from his prospecting operations. This the sergeant obligingly did."[47] Because this type of coordinated solution benefited both the company and the state, Portuguese military commanders typically cooperated, even if they often took these opportunities to remind Diamang that it was an invited guest in the colony. As the Portuguese forces gradually eroded local, armed resistance, by the late 1920s martial campaigns were rarely necessary.[48] By 1938, Diamang's venerable director, Ernesto de Vilhena, could boast to Salazar that, due to the company's efforts, "there exists today, in the entire [Lunda] district, which is almost twice the size of continental Portugal, only one military unit."[49]

The next significant moment of cooperation on security matters between the company and the colonial state occurred in the wake of Congolese independence from Belgium in 1960 and the nationalist uprising in Angola the following year. Even before these events, though, de Vilhena and Salazar had been concerned about the potential for "communist infiltration" in Lunda. Consequently, in 1960 Diamang and the government had collaborated to construct a new road, complete with three new metal bridges (all at the company's expense), "with the purpose of rapidly evacuating the white population—women and children—from the entire operational zone in the event of an [Angolan nationalist] invasion from the Congo."[50] As the events of 1960 in the Congo unfolded, followed by the Angolan insurrection in February 1961, Diamang implored the colonial army to maintain regional stability so as to both minimize disruptions to recruitment and protect potential targets, such as the company's massive hydroelectric power station in Dundo, which powered both the city and several mining operations further afield. While the military obliged (and Diamang paid), the company also created its own defense outfit, the Corpo de Voluntários, composed entirely of white employees.[51] Although the photograph reveals a well-equipped outfit, it's unclear how well these moonlighting engineers and staffers might have fared in actual combat.

In preparing to defend Diamang, both the company and the colonial state were convinced that the Angolan nationalist movement most likely to strike was Holden Roberto's Congo-based União das Populações de Angola (UPA).[52] For this reason, Portugal's feared Polícia Internacional e da Defesa do Estado (PIDE) arrived in the area in 1963 and immediately began cooperating with Diamang's own intelligence division, the Centro de Informação e Diligencias (CID). As Costa Chicungo, a former employee, recalled, "The PIDE and Diamang's secret police began paying much more attention to what people were doing and saying. It was very dangerous to speak openly about the political situation, the war, etc., especially because you did not know who was an informer and who was not."[53] Although Diamang lost a small number of employees who fled to the Congo to join UPA, including some former staff members who subsequently assumed high-level positions in the revolutionary organization, the attack never came. Instead, UPA focused on areas further west to make incursions into Angola, forgoing the riches in the Lunda soil that Jonas Savimbi's União Nacional para a Independência Total de Angola (UNITA) movement would later access and sell as "blood diamonds" during the protracted Angolan civil war (1975–2002).[54]

It is beyond the scope of this study to determine why UPA opted not to molest Diamang's concessionary era. But, local attitudes may well have played a role in Lunda's disquiet during this otherwise tumultuous period. A former Diamang security chief, Leonardo Manuel Judas Chagas, offered several reasons for the lack of unrest. "The local population didn't provide support to the guerrillas because of the company support they were receiving, so when people did try to come into area to 'cause trouble' no one would offer them support. . . . Also, the company stores were very popular and they sold things very cheaply to everyone—black and white—because Diamang didn't have to pay taxes. One could find lots of products in them . . . (even) from Communist countries that Portugal was ignoring. But Diamang sought out these products; for example, shoes from Czechoslovakia and items from China."[55] This decision constitutes a rare instance in which Diamang and the state diverged in a matter of international policy and security. Another explanation for the quietude was anecdotally offered to me by a Lunda resident who grew up there in the 1950s: "Many Angolans didn't really know anything about the war because the movements were not active in Lunda. It all seemed very far away. In fact, I didn't know who Agostinho Neto [the Angolan revolutionary who would go on to become the independent nation's first president] was until I was in school at fifteen years old and the Portuguese teacher started calling Neto a 'bandit, terrorist and drunk'!" Even armed with this knowledge, though, most Africans had no difficulty reconciling their employment with Diamang with the efforts of the nationalist movements. When I posed this possible dilemma to José Silva, he immediately responded: "I had no internal confusion."[56]

SUCCESSES AND SETBACKS: THE EARLY YEARS OF FORMAL MINING OPERATIONS

Upon launching mining operations, Diamang faced serious challenges, including a local population reluctant to engage with it, minimal infrastructure, and nationality-based tension among its white administrative and technical staff. Yet within a reasonably short period the company had succeeded in overcoming these operational impediments, in part because of the assistance provided by Forminière, its established counterpart operating across the proximate Belgian Congolese border. Indeed, although the colonial state provided Diamang with the regional security the company required, Forminière helped the Angolan enterprise during its incipient period in ways that the strapped and overextended Portuguese state simply could not.

Forminière's assistance to Diamang during the latter's tenuous early days included the provision of vital technical and administrative support, safe and efficient supply lines, and even help in repressing the regional commerce in illicit diamonds. A description by Parkinson, who had initially worked for Forminière, of the early years of Diamang's operations highlights the collaborative and even seamless nature of these neighboring firms. "In the early 1920s . . . the [prospecting team] personnel was American and consisted mostly of men who had served their early apprenticeship in developing the Congo fields. . . . Though we had signed on with Forminière, we realized later that we had more or less been leased out or loaned to Diamang. . . . Relations between the two companies were cordial. During the initial phase, Diamang utilized the services of Forminière, both technically and administratively."[57] Equally important was the riverine route to the Atlantic coast that Forminière offered, which was essential for transporting Angolan diamonds out to Europe and supplies and European personnel into Lunda. Not until the mid-1920s did a comparable road network exist in Lunda—constructed largely, and at great cost, by Diamang—connecting the company to the rest of the colony and, thus, the outside world. Going forward, Forminière and Diamang cooperated on several fronts, though by the late 1930s, buoyed by burgeoning profits, the younger diamond enterprise had already assumed the senior position in this relationship.[58]

Forminière also indirectly assisted its Angolan counterpart by having trained on its own mines ethnic (Ba)Luba, who very early on began appearing in Lunda.[59] Although the numbers of these experienced laborers at Diamang were small, and diminished over time, they were, as one company official asserted, "at that time, precious auxiliaries, as they had worked for Forminière and thus had knowledge of diamond mining."[60]

Irrespective of the invaluable experience that most Luba employees possessed, their mere presence was helpful, as procuring indigenous labor for Diamang was initially reasonably difficult. A company report described the labor crisis during these early days, in which African workers numbered only in the hundreds: "The population of Lunda, scarce and without any work habits, and . . . in the early years still hostile to contact with the white, did not permit us to recruit a large quantity of manpower that, from the beginning, was recognized as indispensable to the development of mining operations."[61] Control of the once-lucrative rubber trade, and the possession of firearms facilitated by commerce in this commodity, had generated a type of enduring pride among local communities, rendering the prospect of service to either the Portuguese state or a private entity highly objectionable. Local armed resistance, ongoing military campaigns, and the consequent displacement of portions of the already scarce local population all acted as further impediments to the company's efforts to gain access to these vital human resources, leading ultimately to the implementation of the coercive contratado system.

Animosity between Portuguese and "foreign" white employees also impeded operations in Diamang's early years, with the former acutely bitter that they felt like second-class citizens in their own colony—the irony completely lost on them. Although the company's operational field was in Angola and its headquarters in Lisbon, few Portuguese served on early iterations of its white staff. In 1919, for example, owing to the fact that few Portuguese possessed the necessary technical expertise to occupy skilled positions, the majority of Diamang's foreign staff was American, and the dominant language was English. Two years later, the situation was largely the same, as although the number of Portuguese had risen to sixteen, there were also twenty-two Americans, seventeen Belgians, nine Boers (Afrikaner South Africans), seven English, one Luxembourgian, and one Norwegian.[62] Reports of rifts within Diamang's white staff, including one incident in which African servants had allegedly proclaimed that "the Portuguese were 'nobodies' and that the Americans were in charge in this land," ultimately reached an indignant Norton de Matos. In response, the high commissioner issued a series of decrees that set in motion a process of "Portuguesization" that saw the percentages of Portuguese on staff reach 50 percent by the late 1920s and over 90 percent by 1945.[63] Over time, increasing technical, geological, and operational expertise within the Portuguese staff facilitated this process, helping to create a vibrant "Lusitanian" community in Lunda that informants still remember as a truly halcyon place.

⌒

Even after prospectors verified the existence of exploitable diamond fields in Lunda, Diamang's eventual prominence as Angola's leading revenue generator was neither certain nor inevitable. Before formal diamond mining began, clashes between the Portuguese military and well-armed local communities created an environment that was not conducive to Diamang's ambitions. Moreover, even those local residents who were neither engaged in, nor had relocated away from, these conflicts had little interest in working for the company due to a tradition of self-dependence predicated on the regional commerce in slaves, ivory, and, most recently, rubber. After the state had successfully "pacified" the local population, Diamang turned to it for help in procuring laborers, initiating what would be an extended relationship with the colonial administration that, although strained at times, ultimately served both parties well.

Given the formidable obstacles to commencing mining operations in Lunda, it is remarkable that the company was able to extract even the modest 4,110 carats that it did in its first year of operations. Yet without the current of African laborers from both the Belgian Congo and, in greater numbers, from Angola itself, who first trickled and then increasingly flowed onto the company's mines, even this output would have been unattainable. In the ensuing chapters, starting with the subsequent one outlining recruitment practices and experiences, I follow these laborers and their family members as they traveled from their respective villages to the mines, and then back home again.

3 ∽ The Recruitment Process, 1921–75

In terms of recruiting, all of the state's help is necessary . . . because it has been long since proven that it is rare for an *indígena* ["native"] to offer his services voluntarily, even from the regions pacified long ago, and even less so from those areas in Lunda which were more recently subdued.

— *Diamang Report from 1929, prior to the improvement of conditions on its mines*

I was sixteen when I went. I was not scared because it was my duty . . . even as the nephew of the *soba* [traditional authority]. I had many images of and ideas about the conditions and life on the mines because of all the stories that had been brought back. But again, I did not weigh them too heavily because it was an obligation to go, and so many before me had done it.

— *Former Diamang* contratado, *Paulino, who first traveled to Diamang in 1962*

ALTHOUGH FROM A VERY early stage company officials envisaged a completely volunteer indigenous workforce, until the mid-1960s this objective was unachievable. Before this time, Diamang was repeatedly compelled to sacrifice its long-term labor stabilization goals in the name of immediate manpower requirements, that is, profits. Fortunately for the company, the frugalities of Portugal's colonial project meant that it could rely on forced labor, or *shibalo*, to help staff its mines. The far-reaching state-company agreement of 1921, which required local colonial administrators to procure laborers for Diamang and extended the coercive contratado labor regime to Lunda, greatly assisted the company in meeting its labor requirements until the cessation of the scheme in the mid-1960s. However, even this aggressive forced labor regime could not fully satisfy Diamang's unflagging demand for workers. Because of both the company's burgeoning manpower requirements and the limited success that the enterprise had in attracting African employees from beyond its exclusive labor reserve, Diamang officials incessantly strove to develop a substantial volunteer, or *voluntário*, regional workforce. The fact that it was much less expensive to recruit voluntários than to procure forced laborers (*contratados*) further motivated officials.

Because Diamang's African employees were not expendable the way that they were on mines elsewhere on the continent, after a challenging initial decade and a half of relations with its indigenous labor force Diamang pragmatically adjusted its staffing strategy. From the 1930s forward, as part of its paternalistic approach, the company steadily improved living and working conditions for its African employees, which, in turn, prompted increasing numbers of Lunda residents to willingly engage with Diamang, even if others continued to be compelled to offer their labor.

Prior to this host of service upgrades, Lunda residents had scant interest in tendering their labor to the enterprise. Even a public display of power in 1919, in which Diamang officials paid the newly imposed colonial "hut tax" for the *entire* African population of Lunda in front of regional sobas to publicly signal a shift in the local field of power, did little to convince residents to engage with the company.[1] Following the local installation of shibalo two years later, many residents avoided recruitment by either relocating ahead of the arrival of state labor recruiters (*cipaios*) in their villages or deserting at some point during the journey to the mines. Some of these "fugitives" stayed away only temporarily and eventually returned to their villages, risking capture. Others, however, found permanent refuge in the nearby bush, across the proximate and porous (Belgian) Congo border, or even further abroad.

As time passed, however, most Lunda residents determined that they were inescapably situated in a relationship with the diamond company. Consequently, residents increasingly volunteered their services or simply waited for the arrival of cipaios to escort them to the mines as (forcibly) contracted workers, while their sobas looked on, helplessly. Paulino's testimony, excerpted above, is representative of this exact scenario. Yet even these coerced recruits were able to shape this initial portion of the extended labor process by exhibiting the same type of professionalism that characterized their comportment once on the mines. In exchange for continually improving living and working conditions at Diamang, workers reciprocally eschewed preemptive relocation or desertion strategies in favor of an array of less drastic undertakings, including having family members accompany them to the mines, sharing tasks along the way, and insisting on reasonable treatment during these journeys.

In this chapter, I explore the recruitment process and the changing ways that Lunda residents both experienced and creatively responded to shibalo. I also examine Diamang's shifting labor procurement strategies and policies, beginning with the 1921 state-company agreement, through the dismantlement of shibalo in the mid-1960s, to the eventual conclusion of mining operations in 1975. This examination follows contratado recruits and accompanying family members as they entered the extended labor process, traveling from their home villages to local colonial outposts (*postos*), and finally on to Diamang's headquarters in Dundo before being deployed to the mines. I also consider the ways that voluntários engaged with Diamang and explore the divergent experiences that contratado and voluntário recruits had.

THE GEOGRAPHIC SCOPE OF DIAMANG'S LABOR PROCUREMENT

Until 1921, Africans residing beyond Diamang's immediate area of operations most likely knew little about the enterprise's activities. Up to that point, Diamang had lacked the wherewithal to expand its recruiting efforts much beyond its mining environs. This scenario changed, however, with the landmark 1921 agreement, which committed the Angolan government "to provide all possible support for the recruitment of the indigenous personnel necessary for the intense mining of the diamond beds."[2]

Not content with its exclusive recruitment zone and emboldened by the state's readiness to furnish laborers, Diamang cast its gaze not only on the local Chokwe population in Lunda, but also on areas immediately west (Malange) and south (Moxico) of its concession for desirable recruits. Company officials deemed these regions of greater population density as vital supplements to Lunda's relatively few inhabitants.[3]

Diamang officials also believed that these areas would deliver "higher quality" workers than the local Chokwe population could offer.

As early as 1924, for example, a company official opined that "among the labor force, Chokwe workers . . . demonstrate the lowest levels of physical capacity and civilization."[4] Not until the 1970s did officials begin to soften their stance toward Chokwe laborers. In 1971, for example, Diamang's director general, João Bexigo, admitted that "the company was wrong to think for so long that one laborer from outside of Lunda [i.e., non-Chokwe] was equivalent to two or three native to this region. We are now truly determined to actively overcome the opinion that Chokwe are averse to work."[5] The company's increasingly relaxed sentiments concerning workers' provenances is reflected in testimony from Joaquim António Issuamo, who was employed at Diamang in the 1970s. "Diamang . . . did not care where workers came from; all it wanted was performance and results on the job."[6]

Over time, Diamang's geographical recruitment foci remained relatively consistent, primarily consisting of Lunda and a series of immediately adjacent areas, such as Songo (Malange) and Moxico. Notable exceptions also existed, though, such as Diamang's efforts to recruit "external" groups, including Kwanhama pastoralists from Angola's shared border with Southwest Africa (Namibia); residents of the colony's populous central highlands who were meant to act as intra-Angolan "colonizers"; and even Mucubais from southern Angola who, based on a "defiant" past, were seized by the Portuguese during the early 1940s and relocated to Diamang as "prisoners of war."[7]

In addition, ethnic Baluba regularly crossed the proximate (Belgian) Congolese border seeking work with Diamang and, thus, constituted a much more immediate and consistent source of "external" laborers. In the first two decades of operations, Diamang came to depend heavily on these experienced mine workers, using them primarily as mine overseers. As intended, these prized positions prompted additional Baluba to emigrate. Over time, however, the growing ranks of experienced Angolan laborers and colonial legislation that targeted foreign workers served to reduce their significance.

Regardless of Diamang's grandiose recruitment ambitions, it was largely frustrated in attracting populations beyond Lunda's borders. Even as the company improved working conditions, wages remained low by regional standards. Moreover, foreign, migrant laborers weren't poised to take advantage of local, reciprocal benefits, such as Diamang's provision of free mobile health care.[8] Consequently, Lunda was the provenance of over 80 percent of the African labor force, and most of these employees were ethnic Chokwe. In turn, this ethnic and regional homogeneity greatly contributed to the stability that pervaded Diamang's mines and encampments.

PREEMPTIVELY AVOIDING THE FORCED LABOR REGIME

Well before the tumultuous events of the 1960s, the entire colony of Angola, including Lunda, was hemorrhaging potential laborers. These individuals, or even entire communities, were fleeing not only the manual labor associated with shibalo but also the violent recruitment process itself. Given the colonial state's ongoing dependence on shibalo, the United Nations estimated that by 1954 over half a million Angolans were either temporarily or permanently living beyond its borders.[9] In 1961, Henrique Galvão, a former Angolan administrator-cum-political agitator, cautioned that, "Only the elderly, the children and the feeble remain behind. . . . Angola is fast approaching a demographic—and, therefore, an economic—catastrophe."[10]

Until the mid-1930s, Lunda was no exception to this phenomenon. To avoid the forced labor regime, many local residents initially relocated either within or beyond Diamang's recruitment zones before the arrival of cipaio recruiters in their respective villages. For example, during that decade, a regional administrator of a favored Diamang recruiting area lamented that "my predecessor . . . managed to perpetrate [such] violence and coercion against the indígenas that it has not failed to bring about tragic consequences. This deplorable situation spawned terror among the indigenous populations; the census . . . shows alarmingly decreased numbers, reducing the number of indígenas counted in 1931 by 50 percent."[11] Both company and colonial officials provided a list of foreign and domestic factors that were allegedly responsible for "pulling" these residents away from Lunda, but a general unwillingness to work for Diamang best explains this early response to the forced labor regime.[12]

For those individuals, groups, or even entire communities who opted to flee, the (Belgian) Congo border offered a proximate escape. As conditions in the neighboring colony were arguably even harsher than Lunda's, though, over time this alternative became decreasingly attractive. Only in the 1960s, following the independence of the Congo did this option begin to regain its earlier appeal.[13] Miudo Rafael, a former laborer, explained how residents strategically used the border during this era. "When we heard the cipaios were coming [to recruit], we would flee toward the nearby Congo border and would stay in villages near or over the frontier for roughly two weeks until the cipaios had left, and then would return to our villages. I did this, too."[14] Preemptive flight also gradually became a smaller-scale endeavor, as the colonial state and company were increasingly loath to allow mass relocations. Furthermore, most regional residents simply "needed" the company to survive. Although Costa Chicungo's choice of words is perhaps undue, his testimony reflects this sentiment well: "Most people would not run away before or after being chosen because you were 'lucky' to work in the mines, as you needed your salary to pay taxes."[15] Given these financial responsibilities, and against the backdrop of continually improving conditions both on the mines and during the recruitment process, preemptive relocation virtually disappeared over time.

DIVERGENT ENGAGEMENT PROCESSES, SIMILAR RESULTS: CONTRATADOS AND VOLUNTÁRIOS

The contratados and voluntários who constituted the company's African labor force had vastly different experiences before arriving on Diamang's mines. For contratados, the engagement process began when their sobas offered them to cipaios, who would then lead ever-expanding convoys of recruits and any accompanying wives and children from villages to regional postos.[16] At the postos, local colonial administrators, or *chefes de posto*, either in conjunction with company agents or alone, obliged these recruits to enter into contracts with Diamang, thereby legalizing the arrangement. While at the postos, recruits performed a range of tasks for chefes, who typically oversaw and managed the smallest of these colonial administrative units like personal fiefdoms. From the postos, contratados next set out on foot for Dundo and eventually to their future work sites, a trek that before the introduction of mechanized transport saw some recruits travel more than 1,000 kilometers. These long walks predictably drained the contratados, leaving them weak and in poor condition on arrival in Dundo. Diamang even rejected some recruits due to the debilitating effects of these treks. Only from 1948 did the company provide truck transport for coerced recruits, alleviating this aspect of the engagement process.

Meanwhile, most of the voluntários were men who willingly sought employment with Diamang and who resided within reasonable proximity of the mines (as most returned to their homes each evening following the conclusion of their shifts), and thus this category is also crudely geographical in nature. Another route to attaining voluntário status was for a contratado to choose to remain with Diamang after the completion of his contract, though voluntários of this sort were numerically inferior to the large volume of regional residents who freely sought employment with the company.

CONTRATADOS

Whether contratado recruits set out alone or with family members, their paths to the mines were often as circuitous as they were arduous. The journey began with the arrival of cipaios in recruits' villages and included subsequent stops for processing in both regional postos and Dundo before they finally reached the mines. Although this process could be completed in a few days, before mechanized transport it more commonly took weeks or even months. Moreover, time spent in both the posto and Dundo almost always included daily tasks. Between gathering firewood and cooking during the treks from villages to postos and the variety of work performed while stationed at the posto and in Dundo, the labor process for contratado recruits commenced well before they helped to extract their first diamonds from the Lunda soil.

To escape this fate, some recruits elected to desert during the march from their villages to the postos, while stationed at these outposts, or later, on the way to Dundo, and sought to avoid recapture either within or beyond Angola's borders. In the early years of the contratado regime, these evasive measures involved reasonably low levels of risk. From the 1930s on, however, improved vigilance methods, in conjunction with African laborers' increasingly professional approach and the significant improvements to working and living conditions that the company made, greatly reduced the frequency with which recruits sought to avoid employment at Diamang. Instead, conscripts typically pursued much less dramatic measures, creatively coping with the challenges associated with recruitment, but ultimately opting to remain in the labor regime, rather than abandon it.

Fighting over a Woman's Place: Diamang's Spousal Accompaniment Policies and State Obstruction

One of the most significant divergences in the engagement process for contratados and voluntários was Diamang's desire that spouses and children willingly accompany coerced recruits to the mines. Like their counterparts on the copper mines of Northern Rhodesia (Zambia), Diamang officials displayed great enthusiasm for the relocation of families — or partial families for men with multiple wives — to its mines. Administrators in both settings determined that workers accompanied by family members deserted less frequently, were generally more "stable" in mine encampments, and were more likely to remain on after they had completed their contracts. As Lisa Lindsay has contended, labor stabilization endeavors across colonial Africa were intended to position workers and their families in environments where the labor force could be replenished and a new generation reproduced, while separated from the rural family and other "disorderly" elements.[17] Over time, Diamang proved reasonably successful in its efforts to have wives and family members relocate with contratados and thereby (socially) produce and reproduce. However, obstructive local colonial administrators in Diamang's early days and ongoing, low retention rates of contratados on completion of their contracts frustrated the enterprise's efforts to create a fixed African labor force.[18]

From as early as 1922, Diamang was optimistic about the prospects for spousal accompaniment. Diamang's director, Ernesto de Vilhena, expressed the company's sentiments in a 1929 letter to the governor of Lunda, Vasco Lopes Alves: "For a long time, we have sought by all means to induce the contratado to have his wife accompany him, with the possibility of procreating while on the mines and also because Diamang would be receiving a man who will get on well with his job. If he brings his wife . . . he is a probable future settler of the region who will remain with the company voluntarily."[19]

During the 1920s, the company introduced a series of measures to encourage exactly this scenario. In 1922, for example, Diamang was offering ten escudos to sobas for each wife who accompanied and remained with her contracted husband.[20] In 1925, mimicking approaches used in Northern Rhodesia, the company began offering women a half-ration for the journey and, in 1929, a *pano* (fabric used as a wraparound garment) upon arrival in Dundo. In 1930 and 1945, respectively, Diamang enhanced these incentives to 60 percent and an additional pano upon the conclusion of the contract. By the 1940s, in addition to the above benefits, the company was also granting women a small monetary bonus of 40 *angolars* on their return from the mines.[21]

Although these measures were reasonably successful, local colonial officials occasionally undermined Diamang's spousal accompaniment objectives. For roughly the first three decades of the contratado regime, some chefes do posto contravened a company directive by turning spouses away upon their arrival at postos. Chefes had myriad reasons for this defiance, but their primary motivation was that Diamang's goal of permanently relocating families to the mines would deprive them of their future tax base, from which they derived a commission. Indeed, even when recruits with multiple wives explained that one (or multiple) nonaccompanying wives would act as anchors, thereby ensuring their return, chefes often remained unmoved. A 1945 Diamang document discusses this type of obstruction and reveals the aggravation it generated: "It will not be possible to increase the number of accompanying wives without the help of the [regional] authorities, who currently do not provide it, either through incompetence, or even bad will. . . . The true sense of the indigenous policy escapes the majority of these functionaries, who feel that if women follow their husbands to the mines they will not return to reestablish themselves in the area . . . with a resultant diminution in the [amount of] local tax collected."[22] Even company blandishments, including stipends, cigars, and whiskey, which it allotted to chefes based *solely* on the numbers of accompanying wives emanating from individual postos, appear to have done little to alter these administrators' attitudes. Only after the ascension of Diamang's profits and corresponding sway over the colonial state could it ensure that chefes complied by arranging for any uncooperative administrators to suffer professional censure, an unwanted transfer, or even dismissal.

The company also regularly accused senior colonial officials of diminishing the numbers of accompanying wives by reducing the duration of labor contracts in Angola. These legislative measures were intended to facilitate procreation (to create future laborers) in rural settings by reducing the time contracted husbands spent away from their wives and were, therefore, antithetical to Diamang's labor stabilization strategy. Rows over contract length—which vacillated between six, twelve, eighteen, and twenty-four months—erupted openly in 1928, 1938, 1952, and 1956, with Diamang arguing that longer contracts would translate to increased numbers of accompanying wives, thereby creating more opportunities for procreation (on the mines).[23] The state, however, effectively countered that birthrate statistics undermined the company's assertions and retorted that although "the entire period of the contract is completely lost for purposes of procreation when the contratado is not accompanied by his wife, neither can it be said—even when the contract was for eighteen months and recruits were accompanied by their wives—that they encountered optimum conditions . . . for procreation. The instability and provisionality of the life that they . . . lead is sufficient to ensure that it does not happen."[24] Diamang's inability to convince large numbers of married couples to stay on after husbands had fulfilled their contractual obligations only bolstered the state's contentions, while company officials struggled to marshal evidence to counter even the suspect claims proffered by the government.

A Strategic Decision: Spousal Accompaniment from the Contratado's Perspective

Although Diamang actively encouraged wives to accompany their husbands, the company always left this decision in recruits' hands. Deliberation for husbands and wives revolved around several factors, including any first- or secondhand knowledge of life on the mines, the number of wives a recruit had, and the distance to Dundo. In practice, extended families also participated, helping the couple reach a collective decision before the arrival of the cipaios.[25] Everyone involved considered this social reproduction strategy with the utmost gravity, understanding its importance to the perpetuation of the family and, by extension, the community. The experience of Janette Pedro, who first left for the mines in 1960, was typical: "Everyone was involved—my family, me, my husband—to decide whether I should go or not. Actually, I went to my husband's family, and they gave their opinion, but then I went to my family, and they made the final decision. But this was something that we had talked about before it actually happened. So we had anticipated it and planned for my departure."[26] Other times, though, these inclusive processes produced different results. António Muiege explained that "when I left for the mines [in 1960] I was married, but the family of my wife forbade me from taking her, fearing problems she might have there."[27] In all cases, sobas appear to have respected couples' decisions but also took these opportunities to remind wives of their conjugal duties. As Mulevana Camachele explained, "The soba respected my and my husband's decision to have me go to the mine. The soba said that the wife's responsibility was to her husband so he respected our decision. Others would fill in while I was absent [from the village]."[28]

The most important factor influencing this decision was any knowledge about life on the mines that couples and families might already possess. In general, recruits who were fulfilling second, third, or even fourth contracts were more amenable to having their wives join them. Yet even secondhand knowledge usually allayed concerns. Mawassa Mwaninga, who traveled with her husband to Diamang in 1964, explained that "family members of mine had already gone to the mines, so they knew that it was okay and they did not discourage me from going and were not scared about me going. There was a woman in the village who had already gone, too."[29] The ongoing upgrades Diamang was making on the mines also featured in these conversations, helping to expand the numbers of those who knew that the environment was not only tolerable, but constantly improving.

Prior knowledge could, however, also dissuade husbands and wives from traveling together to the mines. Informants' most regularly cited reason for having wives remain in the village was not a "fear of the unknown," but rather an awareness that assertive co-workers or, worse, aggressive African overseers (*capitas*) or European employees might covet their wives. Fernando Tximvula, who began with Diamang in 1965, recalled that "only four out of my contingent of 120 recruits brought their wives. Men were scared to bring them. The capitas were predatory toward workers' wives, but the whites were, too—especially if she was pretty."[30] Mulombe Manuel's wife stayed behind during his first contract in 1960 for similar reasons. "I did not bring my wife during my first contract, but I did on subsequent ones. I had heard accounts from those who had returned after they worked there and I wanted to see for myself first. Based on their accounts, I was told that if I brought a pretty wife, the capitas would have her."[31] Fueled by periodic incidents of sexual abuse on the mines that Diamang managed to greatly reduce over time but struggled to eliminate completely, this form of trepidation persisted throughout the contratado regime, even if much of it was unfounded.

Another factor that influenced spousal accompaniment was whether or not a recruit had multiple wives. Recruits with more than one wife were more likely to depart accompanied by a spouse, confident that the wife, or wives, who stayed behind could tend to the fields and any personal possessions to ensure the social reproduction of the

TABLE 3.1

Percentage of accompanying wives by origin

Year	Company-wide rate	Place of origin	Percentage
1937	25.7%	Camaxilo	5.0%
		Songo	4.6
1949	22.22	Camaxilo	8.66
		Songo	7.94
1952	34.49	Camaxilo	14.61
		Songo	20.30
1956	27.21	Camaxilo	15.69
		Songo	15.53

Sources: Compiled from the following: Diamang, *Nota de informação* (February 22, 1938), 2, MAUC, Folder 86 35°; *Mapa de mulhers que acompanharam os contratados chegados* (1952), MAUC, Folder 86B,6 5°.

family. Writing in 1954, Borges de Sousa, a company official, indicated that "colonial authorities . . . report that, presently, the majority of indígenas accompanied by a wife have more than one and, thus, are able to leave another in the village to take care of their properties and fields."[32] Although de Sousa's assessment is demographically dubious, it does highlight the strategic advantage that men with multiple wives had.

Finally, recruits and their wives also appear to have factored in the sheer distance to Dundo. The data related to accompanying wives are sporadic, but they suggest that wives coming from further away were less inclined to join their husbands. Couples from Camaxilo and Songo, 775 and 745 kilometers, respectively, to the west of Dundo, were indicative of this trend (table 3.1). Only after Diamang introduced mechanical transportation at the end of the 1940s, thereby greatly alleviating the journey to the mines, did the number of accompanying wives from those two areas begin to increase.

It is difficult to discern broad trends concerning accompanying family members, as the company did not keep reliable statistics on this practice until the mid-1930s. Yet table 3.2 offers some insight into recruits' choices.[33]

The relatively consistent percentages of accompanying spouses from the 1930s forward—which coincided with the improving conditions on the mines—as reflected in table 3.2, indicate that this practice was both durable and largely resistant to disruptions, such as the momentous political events of the early 1960s. Meanwhile, the few

TABLE 3.2

**Percentage of wives, sons, and daughters accompanying contratado recruits
(inclusive of unmarried recruits)**

Year	Wives	Sons	Daughters
1927	7.19%	n/a	n/a
1937	25.70	n/a	n/a
1939	27.21	2.17%	16.91%
1940	40.05	1.54	15.65
1941	35.65	.30	12.59
1942	37.80	3.45	22.17
1944	38.39	11.01	24.34
1945	38.74	11.88	17.93
1946	29.77	8.64	17.71
1947	23.82	7.38	22.24
1949	22.22	n/a	n/a
1952	34.49	n/a	n/a
1954	32.65	n/a	n/a
1955	28.77	n/a	n/a
1956	27.21	n/a	n/a
1957	22.05	n/a	n/a
1959	23.60	n/	n/a
1960	27.06	n/a	n/a
1961	35.07	n/a	n/a
1962	34.53	n/a	n/a
1963	31.30	n/a	n/a
1965	28.43	n/a	n/a

years during which the company produced statistics that considered only married contratados (table 3.3) provide a more accurate reflection of the strategic decisions that husbands, wives, and their extended families made. Furthermore, the consistency in spousal accompaniment rates over time, as featured in table 3.2, suggests that the accompaniment figures from table 3.3 can be extrapolated onto earlier decades. In so doing, it would appear that from approximately the mid-1930s to the mid-1960s almost half of all wives of married contratados accompanied their husbands.

When men and women did decide to proceed together to the mines, male informants most often cited their wives' practice of cooking for them, both during the journey and on their arrival on the mines, as

TABLE 3.3

Percentage of married contratados whose wives accompanied them

1952	52.87%
1954	49.30
1955	44.75
1956	42.37
1962	56.20
1963	57.40

Sources: Compiled from the following: José Marques, Chefe de Gabinete (January 7, 1956), MAUC, 86B,6 8°; Direcção Técnica, Resumo dos trabalhos realizados em 1963. Relatório No. 24, Apresentado ao administrador-delegado (October 15, 1964), 47, TT, Folder AOS/CO/UL-8A3, Pt. 1: 1940–64.

the greatest value that this arrangement bestowed. In fact, when I asked Sacabela Sacahiavo if emotional factors had played a role in his wife accompanying him in 1950, he laughed and declared, "No, no, I didn't bring her out of love—only to cook!"[34] Lest Sacabela appear unfeeling, Lina Machamba, who accompanied her husband to Diamang in 1967, explained, "It was not always the case that men could not or did not want to cook for themselves, but because we knew that my husband would have long working days . . . he would have had to have come home from a long day on the mines and then gone to retrieve water and cook. . . . So, I went to help. When we returned to the mines . . . my husband wanted me to accompany him again for the same reasons: his concerns about working, washing, and cooking."[35] Other times, however, male informants recalled just how grateful they were to have been able to spend time with their wives during nonworking hours and indicated that when wives stayed behind, they had profound "*saudades*" for them—a longing for something absent.[36]

Given the considerable amount of strategic deliberation that went into this decision, when chefes prevented wives and/or children from accompanying their spouses, recruits often complained to company officials. For example, on reaching Dundo in 1939, a group of recruits protested that colonial authorities in Minungo had not permitted them to bring roughly forty willing family members; the colonial officials apparently declared that "it was not their [the conscripts'] true desire."[37] Yet because a wife's accompaniment of her recruited spouse constituted a key social reproduction strategy, couples were not always easily thwarted.

Ultimately, the decision concerning spousal accompaniment was a multifactorial one that featured several contextual and personal considerations. Although certain logistical and perceptual factors militated against this arrangement, improving conditions on the mines and the overall familial benefits derived from spousal accompaniment appear to have provided sufficient counterbalances.

Getting to Work: From the Village to the Posto

The arrival of cipaios in regional villages marked the beginning of the coercive engagement process. During the initial decades of shibalo, local residents typically both feared and despised these enforcers, prompting colonial and company officials to express concerns that the vicious acts that cipaios often committed would deleteriously affect recruitment. In 1922, for example, Angola's high commissioner, Norton de Matos, called for the expeditious recruitment of cipaios for Lunda to assist in the procurement of labor, but cautioned that "they ought to receive instruction from [European] police agents and the military and the best care ought to be taken to obtain, via this instruction, the most . . . rigorous discipline, and through constant supervision ensure that they do not commit even the slightest shameful or extortionate act toward the indigenous populations. To end with these humiliations will be the cornerstone of the future occupation of Lunda."[38] Less than a year later, however, a company official was reporting that cipaios were still abusing recruits: "Armed cipaios invaded the indigenous populations, encircling their homes, whipping the sobas and their wives. . . . I saw, during the interrogations I attended, the deep traces of the beatings; the cipaios took everything that was valuable in order to sell it."[39]

During the initial decades of shibalo in Lunda, brutality of this nature was commonplace. However, just as, over time, Diamang could have uncooperative chefes replaced, so too could it have objectionable cipaios removed. Consequently, contratado informants who undertook marches to the postos as far back as the 1940s and 1950s indicated that although cipaios would still occasionally resort to violence if they deemed the pace too slow, they were, otherwise, harmless. Miudo Rafael, who first traveled to the mines in 1962, explained, "We did not hate the cipaios. They were our brothers. We only hated the Portuguese government."[40] Moreover, many female informants, whom cipaios had similarly escorted, indicated that these overseers never threatened or abused them physically or sexually and had, in fact, treated them no differently than male recruits. Lina Machamba, who first traveled to the mines in 1967, offered a droll explanation for this treatment: "I was treated well by the cipaios on the march to the posto because we, the women, cooked for them."[41] Testimony of this nature complicates reductive characterizations of cipaios as "collaborators," suggesting that these individuals were both complex and ambiguous historical actors.[42] With the eventual suppression of the contratado regime in the 1960s, cipaios played a less consistently direct or overt function in the process of procuring workers. However, even in a diminished role they still constituted potentially violent embodiments of state power.

When cipaios appeared in a village, the soba would typically welcome and regale them. Muhetxo Sapelende, first contracted in 1940, explained that when the cipaios arrived, "A fowl would be killed and eaten, and the people who had been ordered by the soba to serve as the village's recruits would be brought before the cipaios, who would then accept or reject these men. . . . Sobas would ask for one or two people from a family; it was possible to lose both brothers — or two out of many — at once. Also, one might be rejected but the other taken. Cipaios would also walk through the village looking for strong men, especially if the soba was not offering particularly strong ones."[43] Thus sobas appear to have been willing to try to retain their stoutest subjects before ultimately deferring to the cipaios' authority, aware that any dissent could result in corporal punishment, imprisonment, or even removal.

Far from the stalwarts against colonial encroachment that sobas had been at the beginning of the twentieth century, by as early as the mid-1920s these newly acquiescent figures had become reliable cogs in the contratado machine. Testimony from António Muiege, who was a soba in the late 1960s in Lunda, underscores these individuals' dependence on Diamang not only for their power, but for the sustenance of their followers: "I would administer the orders I received from the [colonial] administrator in my village. I did not receive a salary . . . but I did receive food, rations, soap, and other items for my village, and then I would distribute them. . . . I had to comply or else I would have been punished. . . . I might lose food for the village, too."[44] Virtually powerless and consequently marginalized in the eyes of their subjects, Lunda residents increasingly disregarded these authorities and rightly recognized Diamang as their route to personal and familial security. As

Luciane Kahanga, a Lunda resident and former employee, succinctly affirmed: "The company controlled everything."[45]

In return for sobas' compliance, Diamang periodically rewarded them with small gifts, including money, alcohol, seeds, cloth, surplus Portuguese military uniforms, and, beginning in the 1960s, even new houses.[46] In 1962, for example, a company report indicated that "during the year, we helped the soba Samalambo in the construction of his house. For the old soba Chico Carreiro . . . we arranged a kitchen and we think in the next year we will construct a house for him as a reward for his loyalty to the white, who he has served since the time of the [initial] occupation of Lunda."[47] Diamang also honored cooperative sobas with portraits hung in the company's Dundo Museum, though officials could also remove these paintings as a means of shaming sobas who either failed to provide sufficient numbers of laborers or were disobedient in some other way.[48]

Operating in this altered field of power, sobas were unable to protect even immediate family members from shibalo.[49] In fact, several informants indicated that they were either the nephews or sons of sobas but still ended up on the mines as contracted laborers. Cipaios usually, however, first selected individuals who had failed to pay their taxes, or those who were, as Fina da Costa, a former contratado's wife, described, "socially inferior and not contributing in the village."[50] Irrespective of an individual's relative value to his soba, the best the headman could typically do was to offer the recruit food for the journey to the posto, as well as some parting advice. Miudo Rafael recalled that, "Before I departed, Cambachicapa, my soba, said to us, 'Thank you. Good luck. You need to finish your contract and do not cause any problems while there. *Boa viagem.*'"[51]

Just as sobas were unable to shield relatives from recruitment, they were also generally unable to protect minors. Angolan colonial administrations produced reams of legislation over time regulating the employment of minors, yet cipaios concerned themselves primarily with a potential recruit's perceived physical capability rather than his chronological age. Informants explained that cipaios determined if they were of sufficient age to pay taxes—and thus work—by lifting their arms to see if any underarm hair was present, an apparent qualifying indication.[52] As the company's goal was always to procure as many laborers as possible, this approach was somewhat practical, as, for example, many of my informants' professed ages were often approximations rather than precise figures.

For those male residents who had been selected as recruits, the prospect of traveling to and working on the mines occupied their thoughts. As time passed, many recruits and accompanying wives already had some personal knowledge of what their existence at Diamang would be like and used this information to help themselves prepare mentally.[53] Even for those who possessed no firsthand knowledge, though, having previously observed laborers from their villages returning home illness- and injury-free helped to temper their concerns. João Muacasso, who first left for Diamang in 1956, explained that "many others before me had been selected from my village . . . and those older than me had worked before me so I had a great deal of knowledge about conditions at the company before I left. I did not fear going to the mines because so many people had done it before me and all had survived, so I knew that I could."[54] Other recruits, however, possessing similar forewarning, were much less sanguine. Mawassa Mwaninga indicated that after learning that her husband had been selected for work in 1964, "I was afraid to go because I had heard stories about how the whites beat the blacks on the mines."[55] Similarly, Mateus Nanto stated that by 1959 "many from my village had gone [to the mines] before me. So, I already knew about the . . . conditions there. I was afraid of going, but it was obligatory, so I had to [go]."[56] In fact, many informants stressed the "obligatory" nature of the engagement with Diamang and indicated that their acceptance of this fate discouraged them from trying to imagine what their experience would be like. In other words, since it was mandatory, there was no use in pondering it any further.

Company reports echo this sense of inevitability and the positive impact it had on staffing levels on the mines. In correspondences with Lisbon, Diamang officials in Dundo regularly acknowledged the local pervasiveness of Africans' engagement with the company. For example, a 1945 report acknowledged that "we continue to obtain labor in Lunda, absorbing practically the entire [capable] male population."[57] Similarly, in 1952, Diamang officials declared, if somewhat hyperbolically, "We have already told the company directors in Lisbon . . . that in Lunda . . . there are practically no inactive indígenas, but that they all work, directly or indirectly for the company."[58]

When it came time to depart from the villages, contratado recruits and any accompanying family members left their homes behind and traveled with cipaios to other villages to collect additional recruits. Cohorts from individual villages typically ranged between three and ten, though it was not uncommon for a lone recruit to set out from a village or for a community to lose up to fifteen members at once, especially if it was a larger settlement. Over the course of the contratado regime, cohort sizes remained relatively constant, suggesting an awareness by officials that it was prudent to spread the recruitment burden across multiple settlements rather than gut individual villages. Although contributions from a single village may have been small, the overall size of a group could stretch from dozens to hundreds. In general, both the size of groups and the frequency with which cipaios collected recruits increased over time as Diamang's operations expanded and its need for laborers grew. Thus, smaller groups from a particular vicinity suggest an increase in the frequency of recruiting forays into that area rather than a reprieve for the local population. Paulino, a former contratado, proclaimed, "By 1961, the only ones left in my village who had not yet worked for Diamang were youths, who would go when they were old and fit enough."[59]

A contratado recruit's journey to his local posto varied considerably depending on how far removed his village was and at what point during the march he joined the group. Regardless, recruits always made this journey on foot, as generally only footpaths linked villages in Lunda to one another. João Muacasso's account of his 1956 trek to the posto in Camaxilo is characteristic of these marches:

> After leaving my village, we went from village to village to collect more contratados. . . . Small villages would be expected to produce about five workers, and larger ones more like ten. There were about twenty women in my first contingent. . . . We were accompanied by three cipaios on this trip. Our march collecting workers lasted for one month! It was about ten kilometers between villages. People brought . . . food with them to eat. The soba gave us food for the trip, and they [sobas] could also be made to provide food for the whole group—and certainly for the cipaios. We would walk at night, and rest in the villages for a little bit each day.[60]

Another former contratado, Rodrigues, had, in 1958, what could also be considered a typical experience, with a notable discrepancy in the time spent en route to the posto: "Two people were forcibly contracted from my village. There were about 140 people in the group, but it took us [a total of] four days of marching to reach this number. Small villages contributed one or two people, larger ones five or ten."[61] In Rodrigues's case, it appears that he had joined the contingent relatively late, and, thus, although the time spent traveling to procure recruits was brief, the group was reasonably large.

Not all recruits were willing to comply with their imposed fate. Especially before the mid-1930s, many of them fled during the march to the posto. For example, Muhetxo Sapelende, who was initially recruited in 1940, recalled: "We had three days of marching; we arrived at the posto on the fourth day. During the march, twenty people fled! The best time to flee was when everyone was resting, as we spread out in the bush to rest."[62] When desertions occurred, cipaios would return to the village of the *fugido* (deserter) and demand a replacement from his soba, which the headman was obliged to furnish. Sapelende recalled that during his march, "Cipaios returned to these villages and got substitutes. . . . The group stayed put until the cipaio returned with them." Other times, the cipaios would actually catch the deserters, especially, according to Luciane Kahanga, a former employee, "if the deserters had no family and were living out in the countryside like goats."[63] However, the successful capture of fugidos was a rare occurrence and the state's preferred method was simply to demand replacements.

As with the initial relinquishment of recruits to cipaios, sobas complied when the latter returned demanding replacements, facing severe repercussions if they failed to cooperate. According to João Muacasso, "The soba would be imprisoned if he refused to provide a replacement—though I never remember hearing of an example of a soba refusing!"[64] Even when sobas did comply (which they almost always did), cipaios often forced them to travel to postos where chefes would then imprison and/or publicly humiliate them. Cognizant that sobas were often the unwitting victims of recruits' desertions, some conscripts consciously eschewed this option. For example, Itela Joaquim boasted that "I could have fled easily," but he did not, "out of respect and concern for my soba, who I knew would be punished if I did."[65] Unfortunately for many regional sobas, not all recruits were as empathetic as Joaquim.

During roughly the first two decades of its operations, Diamang officials exhibited indifference toward the state's punishment of non-compliant sobas—a calculated reticence intended not to retard the marginalization process of these regional traditional authorities—but loudly and repeatedly maligned the practice of requesting substitutes for those recruits who had deserted. According to company officials, recruits who fled should be apprehended and punished, so as to deter subsequent desertions. In return, chefes do posto retorted that they lacked adequate resources, that is, cipaios, to conduct these "seek and return" operations. Further, even when fugitives were captured, Diamang was not assured of reemploying the captives. In 1932, for example, 41 of 43 fugitives re-escaped after they had been apprehended, precipitating significant company ire toward local colonial authorities.[66] Thus, desertion not only enabled recruits to avoid employment with Diamang, but also engendered state-company acrimony. In turn, the consequent state inertia initially contributed to the allure of desertion as an option for recruits—even if their actions merely transferred their plights to unsuspecting replacements and also potentially led to punishments for their soba(s).

Getting to Work: At the Posto

The arrival at the colonial posto was a transformative moment, as groups of "concentrated" recruits, or *concentrados*, formally came under the direct custody of the colonial state—many for the first time. These concentrados then officially transitioned to full-fledged contratados simply by signing contracts, usually in the presence of a Diamang official and the chefe do posto. That virtually all contratados could not read the document that legally bound them to labor for Diamang was inconsequential, as, via word of mouth, they were all well aware that they were entitled to a salary, clothing, and rations, in exchange for six days of labor each week for a set number of months (figure 3.1).

Indicative of the challenging nature of the extended labor process, chefes do posto obliged arriving recruits to perform unremunerated work. Typically, these colonial administrators assigned only men tasks, which mainly consisted of various cleaning and weeding projects in the general vicinity of the posto, though chefes were in no way limited regarding what they could demand. For example, in 1937, Henrique de Sousa Noronha, a Diamang official, supported a chefe's decision to have concentrados at Vila Silva Pôrto work on the local airstrip, as "we do not want lots of idle men."[67] More often, chefes simply required contratados to clean the postos and immediate surroundings or perform personal tasks for them.

It appears that harsh treatment from cipaios and chefes at the postos prompted some recruits to flee, or at least encouraged this decision.

FIGURE 3.1. A 1928 labor contract signed by Diamang and colonial officials, which indicates the name, origins, and compensatory conditions for a recruit hailing from Bailundo, compelling him to work for the company for a period of eighteen months.

For example, a particularly aggressive administrator of Caluango posto in western Lunda, near Camaxilo, helped precipitate a string of desertions in the early 1940s by forcing concentrados to weed the terrain near the Luita River, some days until approximately 7:00 p.m., without distributing any rations.[68] In these instances, chefes would typically send cipaios back to the village from which the fugido(s) had come to demand a replacement and occasionally to punish the soba, just as they did when a recruit deserted during the march to the posto. Testimony from Muatxissengue, a former employee who was first contracted in 1954, provides an example of a development of this nature. "One person fled during the time at the posto—he had courage. The chefe sent cipaios to retrieve his soba and a replacement worker. The soba was still thrown in jail even though he had provided a replacement."[69]

Reflective of the disagreeable conditions on the mines before the 1930s, some recruits during this early period of mining operations pleaded with chefes not to be sent to Dundo (i.e., Diamang). For example, in the late 1920s, at postos on the Malange-Lunda border from which contratados regularly emanated, but to which Diamang did not enjoy exclusive labor access, a popular refrain of recruits waiting to learn to which commercial enterprise they would be sent was: "To anywhere, except Dundo."[70]

After Diamang began using trucks to transport workers to Dundo in 1948, company officials periodically tried to limit the days spent at postos to four by fixing the 23rd of each month as the scheduled pickup date.[71] Yet as long as the arrival of cohorts of recruits to the postos remained unpredictable, recruits typically spent from one to three weeks at the postos. Although the concentrado system didn't always run smoothly, this segment of the journey was not one that informants considered as overly unpleasant. Mateus Nanto's experience at a posto in 1959 was typical:

> When we arrived at the posto, we had to do what the chefe ordered: wash his clothes, go and fetch water, do whatever he wanted, and we stayed there for a week before we were allowed to leave. When we arrived there, the chefe started to arrange food for us, including two kilos of *fuba* [ground manioc], two of fish, etc., but the majority of the food that we ate was food that we had brought with us or that we purchased . . . on the road. Women cooked for their husbands, and if their husbands had friends there, for example from the same village, then they would cook for them too, but if not, single men were on their own.[72]

While concentrated at the posto, recruits slept in a variety of places depending on what accommodations were available—if any. Although many outposts included some form of rudimentary shelter for concentrados, in 1959 and 1961, respectively, the informants Itela Joaquim and Cafololo Muamuiombo indicated that chefes allowed them to stay in nearby villages, so long as they reported for roll-call each morning.[73] In the mid-1960s, following the termination of the state-run shibalo system, Diamang began to build a handful of its own structures for assembled recruits, though these were still inadequate to handle the volume of workers and family members awaiting transport to Dundo. In 1965, for example, following the death of a young boy at a Diamang-constructed center in Caungula, an investigation revealed that there were 587 people assembled there, though the facility's capacity was only 77.[74]

Prior to embarking for Dundo, at times the company also administered a medical exam that was intended to exclude those recruits unfit for mine labor. With this measure, Diamang hoped to avoid paying for "unsatisfactory" recruits' trips to Dundo, where a more comprehensive examination was given, only to have them rejected at that later point. The company first introduced these screening exams in 1938, but administered them only sporadically and cursorily, and thus medical personnel in Dundo continued to reject many would-be laborers. In fact, very few of my informants reported that they had undergone a medical examination, or any type of physical scrutiny at all, at postos, suggesting that Diamang had largely abandoned these exams by the 1950s. For example, Fernando Tximvula, who first transitioned through a posto in 1967, declared that "there was a recruiting posto [for Diamang] where I grew up. Basically, you went there—there was no medical exam—and then waited for the trucks and a capita [African overseer] to show up and take you to Dundo. No matter what, you would be sent to Dundo—unless you were blind!"[75] Even when the company did administer exams, the lack of diligence and concern was apparent. Filipe Saucauenhe, one of only a handful of informants who underwent any type of medical screening at a posto, recalled that "potential workers were weighed and then chosen based upon their weight and their overall physical appearance. Even small and weak men would eventually be chosen if they returned a number of times. I was chosen my first time and I weighed [only] sixty kilos."[76]

Prior to 1948, contratados and any accompanying family members set out on foot from the postos to Dundo, which at times entailed covering distances greater than 1,000 kilometers, while also hauling vital rations. This trek was fraught with challenges, including food shortages, river crossings, exposure to the elements during the rainy season, and, at times, even lion attacks. The journey could take weeks or even months, occasionally producing fatalities and invariably exhausting the trekkers. Therefore, at least until Diamang introduced truck transport, this undertaking most definitively differentiated the initial experiences of contratados and voluntários.

Just before departing from the postos, chefes distributed rations, which contratados were responsible for carrying to Dundo. Although allocations were usually inadequate for the journey, they typically still weighed upward of 25–30 kilograms, and had to be consumed judiciously to ensure that they lasted for the entire trip.[77] Further, in the early years of Diamang's existence, even when company officials acknowledged that standard journey times exceeded a month, they were stintingly reluctant to issue more than thirty days' worth of rations. One official commented in 1926, "Consequently, contratados arrive in Dundo appearing as though they came from the middle of the jungle."[78]

An infamous 1928 journey undertaken by contratados from Bailundo, over 1,000 kilometers away from Dundo, highlights the company's pre-1930s frugality. The trekking party consisted of 550 adults and 142 children, including 250 families, plus one company official, Francisco Moreira de Fonseca Abreu, who chronicled the excursion. Diamang was eager to showcase this mass "internal colonization" and to tout it as a spectacle heralding what it anticipated would be a recurring event (figure 3.2). Instead, the photograph captures well the drudgery and somberness associated with the trek. Soon after departing, Abreu found himself short of rations and acutely hungry. For the first half of the trip, each member of the group—including Abreu—had been granted thirty days of rations, while forty-five days were needed to cover this distance. The second half of the trip totaled another twenty-three days before the ragged party finally stumbled into Dundo, hailed by the diamond enterprise as "the first nucleus of colonization." While Diamang officials exulted, Abreu immediately set about crafting a scathing account of the trek. In it, he admonished the company for failing to understand that the "realities of the trip," which included eleven

FIGURE 3.2. Families from Bailundo traveling to Dundo, 1928. *Source: Science Museum of University of Coimbra.*

days of crossing rivers alone—an endeavor that at times required two days per river—must not be considered as "delays" but rather should be anticipated and compensated for in the provisioning of rations.[79] Although the subsequent embarrassment that this episode generated prompted company officials to boost future food allotments, going forward ration deficiencies continued to periodically cause trekking parties to reach Dundo reporting hunger.

In response to the demanding conditions that the trek presented—and especially during Diamang's early days—many recruits opted to flee during the long march from their respective postos to Dundo. In this early period of the contratado regime, recruits could desert rather easily, as they often traveled to the mines unchaperoned. During the first full decade of the company's operations, for example, records suggest that approximately 10–25 percent of recruits deserted after initially engaging with the company, though on occasion whole groups failed to arrive in Dundo. In turn, these mass desertions generated considerable distress for the company officials and mine bosses who had been counting on these new arrivals. A 1921 letter from Brandão de Mello to Angola's high commissioner reveals both the prevalence of this strategy and the fiscal damage it was causing: "About five months ago I requested 400 men, of which I received only 100, and 22 from another request for 100, the rest having fled. . . . Of two groups of 30 and another of 50, not one worker arrived, and for all of them we had already paid out . . . salaries, as well as rations . . . which are all . . . unrecoverable."[80]

Given this financial outlay, by the late 1920s Diamang began arranging for cipaios or capitas to accompany and watch over its human investments during this segment of the trek. The company also began constructing shelters along some of the more well traveled routes to Dundo at intervals of approximately one hundred kilometers at which recruits were to replenish their supplies.[81] Even more beneficial for recruits was the decision by the enterprise in the 1930s to begin paying subsidies for both the total distance that they covered and the number of days they spent en route. At times, these payments could approximate a month's pay, which went a long way toward mitigating this challenging undertaking.[82] Escalating profits enabled Diamang to fund each of these ameliorative, if pragmatic, initiatives, while the company was also busy steadily improving conditions back on the mines themselves. Collectively, the company's array of upgrades redressed virtually all of the primary impetuses for flight. Consequently, after roughly two decades of rampant desertion, by the latter half of the 1930s the flood of fugidos had tapered off to a trickle, falling to only 2.8 percent by 1937. The introduction of truck transport a decade later further reduced desertions, though even into the 1970s it remained reasonably

FIGURE 3.3. Truck transport of recruits, c. 1948. *Source: Science Museum of University of Coimbra.*

rare for a group to arrive in Dundo without at least one recruit missing (figure 3.3).[83]

Getting to Work: A Temporary Home in the "Garden City"

Whether recruits reached company headquarters in Dundo on foot or by truck, on arrival most entered a space incomparable to any they had witnessed previously. Shortly after Diamang began operations, the company established its headquarters at Dundo, which over time earned a reputation as the "garden city" (though it more closely resembled a small town, with a population of only a few thousand). In 1948, following a visit to Dundo, a reporter wrote in a paean: "Dundo is a beautiful, gardened villa—a true oasis in the desert of Lunda. . . . It is a model city, a magnificent urban center that gives to its inhabitants the well-being of a metropolitan city."[84] Certainly, the scene was overwhelming. As the photograph reveals, Dundo resembled an archetypal postwar American suburb, and, over time, the addition of streetlights, swimming pools, tennis courts, movie theaters, and a zoo bestowed on it an otherworldly character. With its clean and orderly roads and walkways, quiet streets lined with single-family homes and ubiquitous gardens, both Dundo's design and architecture contrasted sharply with locales elsewhere in Angola, and Diamang eagerly showcased it to visitors as evidence of its "civilized and modern" operational ethos.[85]

For most arriving contratados, experiences in this "oasis" were merely utilitarian. Workers underwent a medical exam and then were assigned "acclimatizing labor" intended to prepare them for the work regimen and general pace of the mines, to which they would eventually be deployed. In this sense, Dundo constituted a transitional time and space, full of foreign stimuli and behavioral norms to which novice employees needed to adjust. To hasten this process, many newly arrived laborers interacted with contratados who had just completed their contracts and were transitioning back home, strategically gathering information and advice that would help them once on the mines.

Personnel from the company's Serviços de Saúde (Health Services) visually assessed the newcomers before administering formal exams to screen incoming male recruits for communicable diseases and to disqualify any physically incapable conscripts.[86] Unlike the vetting measures in the field, these procedures were applied universally and were in place from the beginning of the contratado regime. As early as 1922, for example, Health Services personnel were rejecting a significant number of recruits from Songo afflicted with smallpox, while turning away recruits from Malange who were considered to be "very poor . . . and absolutely useless insofar as any kind of labor is concerned."[87] In these two cases, health officials went so far as to recommend the cessation of recruitment in these areas, but mine bosses' insatiable demand for laborers muted their entreaties.

Throughout Diamang's first two decades, monthly and annual reports indicate that Health Services was rejecting approximately 10–15 percent of the contracted laborers who arrived in Dundo (table 3.4).

TABLE 3.4

Percentage of recruits refused in Dundo

Year	Percentage refused
1925	14.25%
1930 (April)	14.26
1933	13.00
1934	9.45
1935 (December)	10.93
1937	8.96

Sources: Compiled from the following: *Contract Men Received during 1925*, MAUC, Folder 86 20°; Quirino Fonseca, *Relatório anual das actividades da companhia contra a doença do sono 1934* (March 2, 1935), MAUC, Folder 86 27°; Diamang, *Relatório de Julho de 1935* (August 9, 1935), MAUC, Folder 86 28°; Zea Bermudez, *Relatório de Dezembro de 1935* (January 9, 1936), MAUC, Folder 86 30°.

Rejection rates generally declined over time, owing to better, if still erratic, screening in the field, but ironically continued to include (pre-1948) many who would have passed the exam had they not just concluded the taxing walk.

Mine bosses who received these medically cleared contratados were not always as convinced as members of the medical staff had been of these recruits' capabilities. An irate correspondence from the supervisor of the Luaco mine group in 1928 provides an example of the disgruntlement that these divergent opinions could generate:

> We received here at Luaco today a group of 67 contratados. On going over these men carefully I find them to be of absolutely the most miserable types ever delivered to this group. . . . I find that only 36 of the 67 are fit to perform a regular task at our mines. Thirty-one of these men are so small and weak they would only be an impediment to others. These 31 men have been placed in a lot by themselves. Since they have been passed by Health Services in Dundo, they have no current illnesses and the Company would probably be wiser to return them at once to their homes.[88]

Given Diamang's unrelenting labor demands, the supervisor's dismay suggests that these contratados must truly have been unsuited for mine labor.

Beginning in 1930, the company introduced screening measures that were intended to minimize the subjectivity that had characterized earlier assessments. The new evaluative technique that Diamang adopted was the anthropometric Pignet index, which considered height, weight, and thorax circumference.[89] After registering these measurements, the index generated a number that categorized workers according to their relative strength, thereby quantifying the screening process. In theory, medical personnel would then reject recruits who had index numbers greater than a specified target figure. Figure 3.4 captures the measurement of incoming recruits and conveys the level of seriousness with which Diamang regarded this process.

For those recruits whom medical personnel did filter out, a return to their posto, or "repatriation," followed. In keeping with Diamang's pragmatic labor policies, by the early 1930s this trip was made in

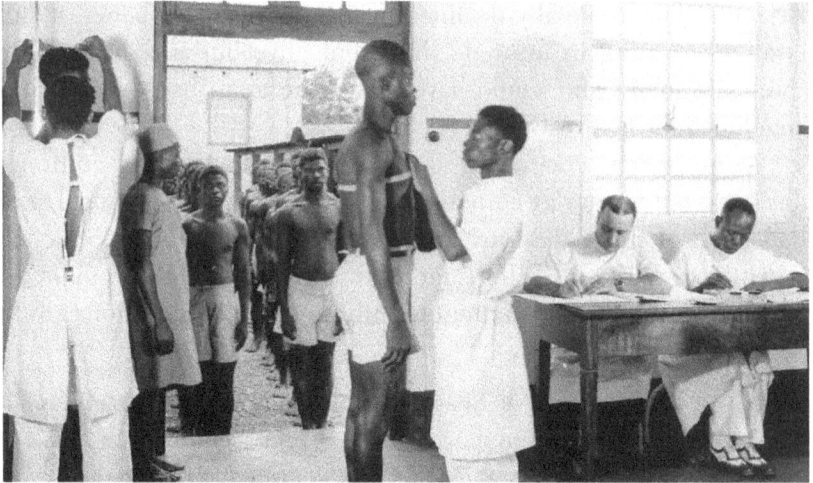

FIGURE 3.4. Measuring incoming recruits in Dundo, 1965. *Source: Science Museum of University of Coimbra.*

company trucks, as the enterprise's directors had deemed that "in cases where workmen are returning to their villages on account of illness, it would give a very bad impression to have sick natives return to their homelands on foot carrying the necessary load."[90] Beyond repatriating ailing workers, Diamang officials had grown increasingly concerned about the negative impact that a recruit who had been rejected due to his "physical incapability" slinking back into his home community might have on recruitment.

Shortly after the institution of the Pignet index, Diamang's Health Services began tracking the results. In the 1930s, company officials set the acceptance/rejection threshold at "33."[91] Yet as table 3.5 shows, only 8.96 percent of contratados were rejected in 1937, while a full 25.9 percent fell into this "rejectable" category and thus should have been excluded. Moreover, the 8.96 percent also included those incoming contratados who may have met index guidelines, but who were rejected for epidemiological reasons.[92] Thus, many of the 25.9 percent of contratados who fell into the "rejectable" realm still found their way onto the company's mines. As this example demonstrates, even after the introduction of the pseudoscientific Pignet index, Diamang officials preoccupied with the sheer number of available bodies could still override recommendations made by medical personnel. Mine managers rationalized their decisions by arguing that these individuals could

TABLE 3.5

Pignet index of 1936–37 incoming contratados

Pignet categorization	1936 Number—Percentage	1937 Number—Percentage
Very strong and good (1–20)	462–8.8%	892–14.9%
Reasonable (21–30)	2423–46.2	2885–48.2
Weak (31–32)	656–12.5	659–11.0
Weak (33–35)	884–16.8	804–13.5
Very weak (36+)	819–15.7	739–12.4
Totals	5,244	5,979

Sources: J. Simões Nevel and Carlos Buso, *Relatório apresentado pelos administradores* (July 31, 1938), 14, TT, Folder AOS/CO/UL-8A. In 1937, the 1,631 contratados from outside Lunda scored the following on the Pignet index: very strong: 509 (31.2%); reasonable: 875 (53.6); weak: 107 (6.6); very weak: 130 (8.6), as opposed to the 4,348 contratados from Lunda, who registered the following: very strong: 383 (8.8%); reasonable: 2010 (46.2); weak: 552 (12.6); very weak: 694 (16.1). Although these figures imply that workers from outside of Lunda were stronger, in practice, the index only poorly measures strength. *Informação relativa à elevação do índice de Pignet* (1939), 4, MAUC, Folder 86 38°.

be given tasks that didn't require significant brawn. Further, these officials contended that steady provisions and hard work would improve the overall constitution of many of these supposedly "inferior" recruits over the course of their tenure with the company.

In 1944, pressure from mine bosses and chefes do posto, who were also straining to produce sufficient numbers of physically capable workers, coupled with the costs of returning rejected recruits to their respective postos, prompted Diamang to elevate the Pignet threshold to 34. This number was a better—though still far from exact—reflection of what the rejection/acceptance figure had, in practice, been heretofore. When combined with better screening in the field, improved rations in Dundo, and a follow-up exam after the conclusion of a contratado's adjustment period, the rate of incoming workers classified as "very weak" workers plummeted to a mere 2.6 percent in 1945.[93]

From the 1950s onward, additional adjustments to the Pignet threshold and the presence in Dundo of more capable medical personnel and increasingly advanced facilities further reduced rejection rates.[94] Testimony from Domingos Cazeweque, a contratado who arrived in Dundo in the late 1950s, reflects this trend: "The second day after we arrived in Dundo we received a medical exam. No one was rejected; even if you had a hernia they would perform an operation on you and then you would go to the mines after you had healed. It was the same with injuries and illnesses."[95] Other informants echoed this testimony, indicating that it was not a matter of *if* you'd be able to proceed to the mines, but simply *when*. Fernando Tximvula, who first traveled through Dundo in 1965, claimed, "In Dundo, even those who failed the exam would be treated and then put to work, not rejected."[96] Internal company reports corroborate this oral testimony, revealing rejection rates of 0.54 percent in 1963 and 0.41 percent in 1965.[97] For contratados, over time merely arriving in Dundo virtually assured them a place on the company's mines.

For the ever-increasing numbers of contratados who passed the medical exam, Dundo became their temporary home as they adjusted to the rhythms and nature of regimented labor. All incoming recruits and family members inhabited the Centro de Trânsito do Dundo (Dundo Transit Center) during this period, sharing the space with contratados who were passing back through Dundo on their way home after having just completed their contracts. Overlap periods with outgoing groups varied, but the official "adjustment period" for incoming recruits vacillated between twenty-one days and a month. In practice, though, these two figures were merely targets. Although time spent in Dundo never exceeded a month, it could last as little as one night if mine supervisors needed personnel urgently enough.

Although company records are silent about contratados' interactions in Dundo, ex-contratado informants spoke particularly fondly about the edifying conversations they had with returning laborers, with whom they commingled each evening. The returning contratados proved to be founts of information for most new recruits, who solicited advice that was even more current, and thus more valuable, than the counsel that they had gathered in now-distant home communities. Lucas Macafuela recalled that during time spent in the Transit Center in 1956, "We requested and received a great deal of advice from these just-finished workers. . . . They stressed how important it was to learn — our jobs and otherwise."[98] Muatxissengue also remembered receiving occupational advice during his time at the center: "We waited in Dundo for about fifteen or twenty days. There, we had an opportunity to talk to those workers on their way out who shared their impressions with us and told us stories about the mines that conveyed to us what we should and shouldn't do there."[99] In both of these examples, the advice given captures the professionalism that Diamang's African laborers were strategically practicing and promoting. Reassured after witnessing almost invariably healthy workers passing back through Dundo, and better informed owing to the updated accounts that these experienced laborers had provided, incoming contratados were now as prepared as they were going to be before their deployment to the mines.

Removed from their respective home communities months earlier, by this point contratados had already endured lengthy treks, presumably unprecedented encounters with medical staff, and perhaps even transport in mechanized vehicles for the first time.[100] Yet, rather than allow these raw recruits to relax while waiting to be sent to the mines, Diamang thrust them and any accompanying wives squarely into the labor process, requiring that they work on gendered teams performing "acclimatizing labor" in and around Dundo. At times, the company excused women from this type of work and, on rare occasions, even men. However, from the early 1920s through the end of the company's operations, most of these transitional workers could count on working in and around Dundo, weeding, cleaning, and washing, six days per week from roughly 7:00 a.m. to 4:00 p.m., with an hour break for lunch. The work itself was not particularly demanding, though, and thus recruits' time there was not overly difficult. Mulevana Camachele, who transitioned through Dundo in 1960, even proclaimed that "life was easier there than it was at home. The work in Dundo was lighter. . . . Life in the village was hard!"[101] Other informants simply recalled their time there dispassionately, perhaps because of the more challenging experiences that they had already endured and/or which lay ahead. Rodrigues even indicated that the work was a welcomed distraction. "We waited in Dundo for one month. . . . We were not afraid of being there . . . because of the tasks we had to perform each day, which kept us busy and our minds off of all that was happening."[102]

When it finally came time to relocate to the mines, workers lined up while a company official counted them off, assigning individuals to this or that mine. For contratados, this method could seem arbitrary or perhaps even procedurally indifferent as, although Diamang did not separate couples or families, once the count for one mine ended and the next began, it was simply a matter of numbers. Thus, when friends or home cohorts did stay together for the duration of their contracts, as many did, their juxtaposition had as much to do with this occurrence as any other factor. João Muacasso, who first experienced this procedure in 1956, described this decisive moment: "I stayed with some people from my original group entirely by luck. The chefe counted off a number necessary for a particular mine and that [here, he used a downward, chopping motion] was where the contingent for that mine ended, and the next person would end up at the next mine, and so on. In this way, I stayed with two people from my village, but others were sent elsewhere."[103] Luciano Xacambala, who arrived in Dundo as a contratado in 1951, recalled, "Three of the people from my village went to one mine and two were sent to another. This [arrangement] was along married/single lines, so that the two who were married went to one [mine], while I went to Mussolegi mine as part of the group of three single workers."[104] Xacambala's experience does not reflect an intention by Diamang officials to group contratados according to marital status, but rather may have been a result of a particular mine needing additional female workers (wives) in either its kitchens or in nearby company plantations or, more likely, was merely coincidental. Beyond periodically experimenting with exclusively contratado or voluntário mines—and in sharp contrast to mining settings elsewhere in Africa—over the course of Diamang's history, company officials demonstrated little interest in socially engineering its mines and encampments along ethnic or any other classifying lines.

VOLUNTÁRIOS

Unlike contratados, who typically came from the periphery of Diamang's recruitment concession, most voluntários lived reasonably close to either Dundo or one of the company's mines. Instead of being rounded up, voluntários freely sought employment or attained this status by willingly remaining with Diamang on completion of their initial contracts.[105] As such, the voluntário category comprised two subgroups: "true volunteers" and "ex-contratado volunteers," which, over time, collectively constituted roughly one-half to two-thirds of the African labor force.[106] Diamang rarely differentiated between these types of voluntary workers in its records, though, so it is difficult to quantify their participation. However, it is safe to say that over time hundreds of thousands of Africans freely engaged with Diamang, even if many of their co-workers were present as a result of coercive measures.[107]

The "true volunteers" who Diamang was able to attract typically experienced an abridged engagement process, traveling from their respective villages straight to an active mine. For this voluntary labor force, distances to the mines had to be traversable in hours—rather than weeks or months—as until the 1940s, virtually all voluntários daily commuted on foot between the mines and their existing homes. Testimony from Costa Chicungo, a former voluntário who began at Diamang in 1952, is exemplary: "When I decided that I wanted to work, I simply walked to the mine and asked for work. I lived in a village very close to N'Zargi mine."[108]

Diamang also expedited the initial engagement process for voluntários by allowing them to enter into short contracts or even to forgo them altogether. As voluntários typically constituted the most proximate layers of potential laborers, Diamang hoped that these individuals would ultimately become "repeat" or even "permanent" volunteers, and, therefore, it adopted a more relaxed approach to their contracts.[109] Consequently, the majority of voluntários engaged with the company via short-term contracts or without one. Although João Muacasso represents an extreme example, his testimony reveals the allure of short-term contracts and the liberty to freely decide whether to (re-)engage, or not: "I would stay in the village once I had returned from the mines. Then, if I had a good friend, I would go back voluntarily to the mines with him. . . . I did this . . . after I had returned home. . . . I spent ten months on the mine the second time. I did this again and again for ten years!"[110] Over time, an increasing number of Lunda residents also worked as informal "day laborers," free to come and go from day to day and receive payment only when they worked. Deque, who first volunteered at Diamang in 1956, explained that "I was a voluntário and went to work . . . on the mines because I was from right near Cassanguidi [mine]. . . . I went alone to the mines to request work. Others in my village worked in the mines and returned home each night because it was so close, only three kilometers. I did not have a contract and started working the same day that I arrived to request work, though I did not work every day."[111] Given voluntários' proximity and the contractual casualness that Diamang exhibited toward them, most of these voluntary laborers started working for the company within a few short hours after requesting employment.

Over time, Diamang's calculated paternalism efficaciously attracted an ever-increasing number of voluntary workers. As the laborers whose provenance was most proximate to Diamang's installations, they were well aware of the improving living and working conditions on the mines and were well positioned to take advantage of local, company-provided services. Moreover, the enterprise also adopted several policies intended to provide incentives to these potential workers. For example, many voluntários were attracted to the lighter jobs that were reserved (though certainly not guaranteed) for Africans who willingly sought employment. Costa Chicungo confirmed that "some villagers would go to the mines voluntarily, so as to avoid becoming a contratado."[112] Diamang officials encouraged this behavior by emphasizing that volunteering represented the first step toward securing coveted occupational postings, such as motorist, that were unattainable for contratados. By the end of the 1960s, the company was also rewarding voluntários with, among other things, elevated wages, "European" meals, and paid days off. Aware that contratados were roughly two and a half times more expensive to recruit and, therefore, employ than voluntários, these supplementary expenses were quite tolerable for company officials, while also proving attractive to potential laborers.[113]

If Diamang's range of incentives appealed to many voluntários, others willingly sought employment with the enterprise primarily for the cash, in the form of wages, that it was offering. For example, Mulombe Manuel, an "ex-contratado," stated that "I returned [willingly] to the mines because I was not earning enough to survive and with the company I would. It was about three to four years before I returned to work. My second contract was voluntary; it was the only company to work for in order to survive."[114] Other voluntários were similarly motivated by the prospect of drawing a regular salary, but primarily only to increase their spending power. For example, Augusto Funete admitted: "I started with the company in 1968. I went voluntarily to the mines, by myself. I knew others from my village who had already worked in the mines, but I did not learn much from them because they apparently did not want others to be informed. . . . They were afraid that if others went that they might not be able to go back. They came back with money and clothes, and that was alluring."[115] Similarly, Domingos Cazeweque boasted that "I had brothers who managed without working for the company. They sold fish, manioc, etc. But, I wasn't envious that they never went to work for the company. They stayed [behind] . . . but I earned more money than them."[116] The low wages on offer at Diamang minimized the overall amount of money circulating in Lunda, yet this oral testimony suggests that Diamang's introduction of cash wages had shaped, at least to some degree, economic relations and strategies in Lunda. It is important to note, however, that a financial motivation to voluntarily seek employment with Diamang was absent from the remainder of the testimony I gathered, suggesting that this impetus was neither pervasive nor should the impact of the regional infusion of cash be overstated.

⮎

Over the course of Diamang's operations, colonial recruitment schemes ensnared hundreds of thousands of African laborers. Reminiscent of the way that slaves were earlier transported through these same lands, cipaios rounded up recruits and compelled them and any accompanying family members to undertake long journeys that ultimately landed them on the company's mines. Entering into this extended labor process was a fate that these contratado recruits had often anticipated and with which they typically, if often resignedly, complied, especially as conditions at Diamang improved and the company mechanized the journey to Dundo. Meanwhile, due to proximity, employees who volunteered their services only rarely endured challenging trips to the mines.

Although most Lunda residents opted not to take flight, in Diamang's early years a steady stream of potential recruits flowed out of their rural villages ahead of the arrival of cipaios. Alternatively, some recruits chose to desert after cipaios rounded them up. Over time, however, conscripts largely abandoned these drastic measures in conjunction with Africans' increasingly professional approach to employment with Diamang. Meanwhile, the diamond enterprise was actively improving conditions both during recruits' journey to Dundo and once on the mines, and was also consolidating regional power and control over local state officials. Thus, even as the company was busy narrowing Lunda residents' labor engagement options, it was simultaneously making its mines a more agreeable place on which to live and work.

In the ensuing chapter, I examine the shifting working conditions, procedures, and policies that Diamang's African employees experienced, as well as the ways that they negotiated this constantly changing labor environment. As full-fledged employees on and around the company's mines, they entered into a series of labor processes that challenged even the sturdiest men and women.

4 ⮑ A Group Effort

The Collaborative Process of Diamond Extraction,
1917–75

There were no machines then. The machines were the people.

> —*Paulo Leão Vega, describing the situation on*
> *Diamang's mines in 1946 when he joined the company*

PAULO LEÃO VEGA WAS RIGHT. Although Diamang significantly improved the conditions on its mines from the 1930s forward, its fiscally motivated reluctance to mechanize meant that until the 1960s production rested almost exclusively on the backs of African workers. Throughout the company's operational existence, virtually all contratados and most voluntário employees prospected for potentially rich diamond deposits in the remote hills and streambeds of Lunda, shoveled for long hours in search of the precious stones, and then transported the gravel to centralized refineries. As part of its ongoing struggle to sufficiently staff its operations, over time Diamang supplemented the adult workforce with young boys who performed many of the same tasks as their elders. Meanwhile, women who accompanied their contracted husbands to the mines rarely remained idle. Many labored long hours in the fields generating foodstuffs for the African workforce, while others served in company mess halls or cleaned the mining encampments.

Although the durable nature of this taxing, manual-labor regime appears to be incongruent with the overall improvements the enterprise was making, the prodigious number of laborers required to realize this operational approach helped foster a regional dependency that was central to Diamang's calculated paternalism. This corporate ploy also included occasional wage increases, though at a pace that rendered the company one of the lowest-paying enterprises in the colony. Yet unlike on South African mines, on which "the constant ability to expand the geographic pool from which the migrants came enabled employers to keep blacks' wages low and almost static in real terms," Diamang justified its wage scale by providing a range of unrivaled services for its employees, including food, housing, and health care, and also the payment of local tax obligations for contratados.[1] The company correctly calculated that the provision of these vital forms of assistance would outweigh the challenging nature of the labor regime. While the low wages deterred Africans from beyond Lunda from relocating to the company's mines, local residents' appreciation of this regional "investment" effectively offset periodic complaints about salary levels, or even the work regimen itself. Modest salaries also minimized the circulation of cash within Lunda, thereby curbing any proletarianization that the impact of a more robust cash economy would have fed.

Diamang's long-standing reliance on manual labor over more costly machinery also helped promote harmony among its African workforce by greatly limiting the creation of skilled or specialized labor positions. The occupational homogenization produced by this thrifty approach precluded, or at least forestalled, the development of a potentially divisive remunerative hierarchy. Even when Diamang pursued mechanization more rigorously in the 1960s, this operational modification generated only a limited number of skilled positions, never constituting more than 10 percent of the overall African workforce.[2]

This chapter explores the changing labor processes in which African men, women, and minors engaged at Diamang and focuses on the uneven impact of the company's gradual efforts to mechanize portions of its operations; the shifting size, composition, and organization of the workforce; the different tasks that Africans performed; daily work schedules; the manner in which African and European overseers supervised the labor force; the occurrence and aftermath of occupational accidents; and worker compensation. It also explores the substantial work of women, whose labor the company rendered invisible, but whose contributions were vital to Diamang's operations and to the social reproduction of their families.

THE GROWTH, NATURE, AND ORGANIZATION
OF THE AFRICAN LABOR FORCE

To staff its expanding operations, Diamang required an ever greater number of African laborers. From the company's inception in 1917 to Angolan independence in 1975 average annual workforce sizes grew dramatically, from approximately 500 to over 27,000. The composition and organization of this labor force stands in sharp contrast to mining workforces elsewhere in southern and central colonial Africa, which were predominantly made up of male laborers and typically organized according to ethnicity.[3] Voluntários generally lived at home and worked either without contracts or with commitments measured in months and, over time, constituted roughly two-thirds of Diamang's African staff.[4] The remainder of the African workforce consisted of contratados and any accompanying family members, who were obliged to commit to work for at least a year. Over time, Diamang increasingly organized its excavation workforce according to an employee's status: contratado or voluntário, with the former performing more of the excavation work. This organizational logic was intended to boost the numbers of Africans who willingly sought work on the mines and, ultimately, to assemble a fully voluntary labor force. Yet in the early decades of mining the difficulty in procuring contratados hindered the application of this labor scheme, while going forward overall manpower shortages inhibited it. During these latter episodes, company officials' determination to meet production targets overrode any concerns related to how attractive Diamang might be for prospective voluntários; where there were openings, bodies were indiscriminately sent to fill them.

THE BURGEONING WORKFORCE

Because Diamang's senior management believed strongly in the correlation between production levels and the number of workers on hand, the enterprise rarely bridled the expansion of its labor force. Even when its relentless push to add additional laborers drove down the quality of incoming recruits, increased the numbers of inexperienced workers, and generated problems integrating so many new workers, Diamang never altered its policy. Only during the initial years of the Second World War, owing to the reduced international demand for diamonds, did workforce sizes briefly stabilize. Other periodic disruptions caused serious alarm among mine managers, but were quickly addressed by pressuring chefes do posto to produce additional recruits. After the

1940s, chefes increasingly turned to minors to satisfy these requests, a practice that company officials condoned. Although more aggressive mechanization would have obviated the need for these younger workers, Diamang's demand for recruits of all ages was unwavering.

In 1918, Diamang's African labor force stood at roughly 895, a portion of which it had inherited from its predecessor, PEMA. A decade later, over 5,000 Africans were working for the company, in great part due to the landmark 1921 contract with the colonial state. Despite this remarkable growth, labor shortages periodically undermined efforts to reach production targets. In 1923, for example, the engineer G. H. Newport reported that "production has dropped during the last two months due to the native labor shortage."[5] A few years later, H. T. Dickinson similarly lamented that "the shortage of native workmen necessitated the concentration of available labor on some mines and the closing down for certain periods of others. . . . I know that in October, Mai Toca mine operated for only ten days, Nzarghi 2 mine, twelve, and Nzarghi 6 mine, seven."[6] As individual mines historically averaged between 100 and 250 workers, even small interruptions in the labor supply could be highly problematic.

When shortages occurred, Diamang typically strove to redress them by either demanding more contratados from existing sources or seeking recruits from new areas. By the 1930s, however, these solutions began to show signs of wear, producing less capable recruits as regional populations became increasingly gutted. Consequently, by the end of the decade, although the number of workmen had risen from roughly 7,200 to over 10,000 during the period from 1935 to 1938, the amount of *cascalho* (diamond-bearing gravel) processed was not reflecting this growth. A 1938 company report attributed this productivity issue, at least in part, to the rapid expansion of the workforce: "One influence on efficiency . . . is the decrease in the quality of labor now being employed as compared with what it was a few years ago. This change was unavoidable as . . . our labor supply areas . . . have been severely strained by our rapidly increasing demands. Directly associated with this influence is the introduction into our mines of many first-term employees. No matter how willing or energetic they are, they cannot achieve the same output as . . . more experienced employees."[7] The report also cited the considerable challenges that introducing so many new laborers into a constantly shifting configuration of mines presented: "It was impossible and hardly unexpected that the recruitment and arrival of workmen from several hundred kilometers away could be exactly synchronized."[8] Thus, during the first two decades of Diamang's operations, expansion-related growing pains occasionally complicated the otherwise reliable correlation between the overall number of laborers and production levels.

By the early 1940s, the company was also increasingly turning to minors, defined officially as either male voluntários or sons of contratados between fourteen and sixteen years old, to expand the workforce.[9] However, as perceived physical capability was more determinative of one's utility to Diamang than age, boys as young as eight were providing service on the mines, and both boys and girls on company-operated plantations.[10] In 1947, the company employed over 400 minors on its mines, while by the 1960s, the number had increased considerably, mirroring the general enlargement of the African labor force.[11] In 1961, for example, there were 1,873 minors "from the region" out of a total of 18,705 such geographically designated workers at Diamang—or about 10 percent, up from the historical average of less than 5 percent.[12]

By the end of the Second World War, the company was annually employing over 15,000 African laborers of all ages. In practice, Diamang's persistent expansion of its workforce diverged from the approaches of mining enterprises elsewhere in the region. For example, after diamond companies in the Belgian Congo launched a mechanization program in 1938, this effort and subsequent others precipitated a 50 percent reduction in the number of African laborers between 1942 and 1958, while productivity rose almost fourfold.[13] During this same period, Diamang officials remained stubbornly content to rely on increased numbers of manual laborers and focused instead on improving the scheduling, allocation, and transference of laborers to minimize the disruptions that the influxes of new laborers generated.[14] Over the course of the 1950s, Diamang increased its African labor force by more than 8,000 workers and by the mid-1960s it was consistently averaging over 25,000 African employees annually. Thus, even as Diamang began mechanizing greater percentages of the core mining functions, it was also aggressively adding manual laborers.

THE COMPOSITION AND ORGANIZATION OF THE WORKFORCE

In Diamang's early years, the company used both contratados and voluntários to perform the arduous tasks that constituted the central mining process and also experimented with wholly contratado or voluntário staffs on individual mines.[15] This trial segregation reflected a company strategy that called for a workforce composed entirely of voluntários, yet the undependable and short-term nature of these workers hampered production and thereby extinguished these designs. L. J. Parkinson wrote to Ernesto de Vilhena in August 1931 to warn him of developments indicative of this pattern. "We have been rather severely handicapped during the past dry season by a shortage of labor, which was brought about by the unstable nature of the voluntários, a great many of whom did not remain at work for more than a few weeks."[16] Indeed, in 1930, contratados constituted almost half of Diamang's labor force (2,539 to 2,564 voluntários).[17] Consequently, over the ensuing years the company pragmatically began launching a wave of paternalistic initiatives to attract voluntários, while also privately admitting that contratados were essential for the manually and numerically intensive excavation and transport tasks if the enterprise was to meet its production targets.[18]

Although never systematically, over time Diamang displayed a tendency to assign tasks according to whether an employee was a contratado or voluntário. A 1946 inspection of Maludi 6 mine confirms this approach to occupational allocation. Of the 83 contratados present, 58 worked shoveling away overburden (nondiamond-bearing topsoil) while another 22 were responsible for removing cascalho (the diamond-bearing gravel found underneath), leaving only three to perform lighter work.[19] A letter from a company official in 1948 was even more revealing. "The voluntários, as a general rule, do not want heavy work on the mines, thus . . . of the approximately 7,000 [Africans] who are utilized in the tasks of removal . . . transport and treatment of cascalho, around 6,000 of them are contratados, while the remaining 1,000 contratados perform lighter services on the mines."[20] António Batista, who worked at Diamang intermittently between 1960 and 1975, summarized his observations of the company's organization of labor: "We worked on the mines together with voluntários, but there was a division of labor, with contratados often getting heavy work and voluntários getting lighter work."[21]

AFRICAN MEN AT WORK: PROSPECTING

The generation of profits at Diamang and the geographical expansion of the company's operations were predicated on prospecting teams' identification of extensive, exploitable diamond clusters in the

Lunda bush. African employees made up most of each prospecting team, which included between five and two hundred workers, one or two white lead prospectors, and, at times, African overseers (*capitas*). Owing to the difficult nature of this work and the extended periods for which teams were away from mining centers, isolated, and with limited supplies, prospecting was the least desirable task in the mining process. Consequently, Diamang increasingly—though never exclusively—began relying on contratados to staff the growing number of prospecting teams. Over time, the company outfitted these teams with more sophisticated equipment, which altered the labor process by reducing the amount of manual labor required to identify deposits, though large numbers of shovelers still featured on each team.[22]

The method of prospecting changed little over the years. If necessary, teams cleared away any obstructive brush before removing samples of earth. Lead prospectors then "spot-gauged" the concentration of diamonds within these samples. As Angola's diamonds are alluvial in nature, these teams traversed the hillsides and streams of Lunda, spreading ever outward from known clusters. Former lead prospector Lute J. Parkinson provides an account of the process as it stood in the 1930s: "We . . . put down exploratory pits at intervals of one or two kilometers. The pits were commonly two or three meters in diameter. Occasionally, an alternative method was employed for taking a small sample, using a hand-operated Empire drill . . . forced into the ground by the weight of four natives standing above the drill on top of a platform."[23] Joaquim António Issuamo, a former prospecting team member, some thirty years later described the nature of the work as he had experienced it, "Teams were usually composed of between six and ten men. We would check [dig and remove a sample] every two meters, so it was very tedious, and if we found something, then we would start checking every meter or half meter, so it took a long time."[24] Both accounts reveal the dull, if challenging, nature of this endeavor.

Originally, African workers removed these soil samples with shovels. Over time, however, Diamang increasingly furnished prospecting teams with updated equipment, such as the Empire drill cited by Parkinson. These new tools improved speculative accuracy, but did little to lighten the workload. Mário Samuhaniquime, who began on a prospecting team in 1969, explained that "the drill was very heavy; it took fifteen people to move it. We had to lug it back to the main roads where trucks could then pick it up."[25] African workers also had to carry the samples away from the dig sites for testing, a task that even into the 1960s and '70s was exclusively performed using baskets. For example, Bartolomeu Lubano, who worked on a prospecting team in the mid-1960s, indicated that "my job was to carry out cascalho from the prospecting site for testing . . . on my head, often about two kilometers."[26] Similarly, Caiombo Jombe, who first joined a prospecting crew in 1964, declared, "My job was to lift and carry out cascalho. . . . We would transport the cascalho on our heads to places where it was tested."[27]

Beyond the difficult nature of the work itself, African laborers disliked the responsibility of constructing their own makeshift accommodations and were especially averse to serving for long periods isolated in the Lunda bush. Samuhaniquime recalled that "the first thing we would do was build an encampment. . . . We would then stay out in the field as long as necessary. It could be months until an area was considered finished. Then we would move to the next area—no break . . . this was our life and job."[28] Parkinson's memoir reveals that conditions were roughly similar thirty years earlier:

> When a payable deposit was encountered, one man might spend several months putting in lines of pits at fairly close intervals in order to prove up the ground. On the other hand, a small centrally located base camp might be occupied only a matter of several weeks while the lone prospector and his native workers traveled daily from it to the pits he was putting in along the ten to fifteen kilometer length of a stream. Sometimes we could work six months in a barren area without finding anything.[29]

Although prospecting was not an inherently dangerous undertaking, because of the objectionable conditions most African employees preferred even the rigors of mine labor to prospecting.

Rather than dramatically improving conditions for prospecting parties, the company increasingly resorted to a reliable source: contratados. Because voluntários who were forced to perform more arduous jobs, such as prospecting, rarely stayed on past the end of their (short) contracts, thereby causing procedural interruptions, contratados were typically assigned these tasks. Company records indicate that early prospecting teams relied heavily on these contracted laborers. For example, a team in 1929 led by a lead prospector named Hervey counted 67 contratados and 21 voluntários.[30] By the early 1950s, however, prospecting teams could be composed completely of contratados. Cognizant of the aversion to prospecting, Diamang officials often appealed to chefes do posto to earmark "obedient and strong" recruits for prospecting duty.[31] Yet indicative of these local colonial administrators' slack approach to recruiting laborers—irrespective of their age—for work at Diamang, both Caiombo Jombe and Bartolomeu Lubano, the former cascalho carriers cited above, were just twelve years old when they joined prospecting teams in 1964 and 1965, respectively. Although Jombe and Lubano may have been "obedient," they were certainly far from "strong."

Despite the problems associated with staffing these teams, Diamang continued to expand its prospecting endeavors both in quantity and in geographical focus. For example, at the end of the Second World War, the company had nine active prospecting crews, all based in Lunda, whereas in 1969, twenty-nine prospecting crews were active in Lunda, two in neighboring Malange, and twelve in Bié, in Angola's central highlands. Only roughly a decade after the outbreak of the Angolan revolution did Diamang scale back prospecting activity, abandoning it completely in 1974. As Pedro Bento Marques, a former prospecting team member, explained, "In 1974, Diamang suspended prospecting activities because by nature you were out somewhere in the bush where there was not an established camp. The company was afraid of attacks, though they never occurred."[32]

AFRICAN MEN AT WORK:
SHOVELING FOR WEALTH

After prospectors had identified payable diamond deposits, African laborers set about excavating them. Workers first removed the top layer of earth: the superficial, nondiamond-bearing "overburden" (or revealingly, in Portuguese, estéril, or infertile land), before digging up the adjacent layer of diamondiferous gravel known as cascalho.[33] The labor process associated with the removal of these two layers varied according to the type of terrain and also changed over time. The introduction of heavy machinery increased the amount of overburden and cascalho that could be removed in a given period, though it never decreased the company's demand for manual laborers. Mechanization initiatives notwithstanding, removing these two layers of earth consumed the efforts of approximately three-quarters of all African employees who worked at Diamang.[34]

Although the excavation of estéril and cascalho constituted more desirable occupational assignments than prospecting, they were the most laborious tasks on the mines, especially during intense heat and inclement weather. Moreover, although African and European overseers decreasingly resorted to corporal abuse to ensure compliance and/or to encourage maximum effort, this form of violence did not entirely disappear until the 1970s. Consequently, Diamang used contratados in these positions when possible, though over time it also deployed many voluntários, especially when shortages of the former occurred.

The fundamental process of removing alluvial diamonds from Lunda's hillsides and riverbeds remained largely manual throughout Diamang's operational era. Over time, both procedural and mechanical measures increased the volume of carats extracted and treated, but manual labor remained central to the overall process.

Indicative of Diamang's ongoing commitment to manual labor, in 1947 Diamang's director, Ernesto de Vilhena, asserted: "We think that the mechanization to be carried out in some of our operations in Lunda should be done gradually and with the utmost prudence. Considering the conditions in which we work there, we could never contemplate, at least in the near future, the economic practicability of a large mechanization scheme. . . . The program to be adopted is a moderate one."[35] In fact, just a month before de Vilhena's expression of these sentiments regarding mechanization, W. A. Odgers, a consulting engineer, had written to the director general: "No large mechanization program is warranted unless there is a very serious probability of a shortage of native labor. . . . I suggest that any mechanization beyond what I previously suggested would prove uneconomical when compared with the cheap native labor available. . . . Due to the comparatively low cost of native labor in Angola I consider that before any major program of mechanization is embarked upon, every attempt should be made to economize wherever it is possible using native labor."[36]

Whether by mechanical or manual means, the first step in the extraction process was always the removal of the unwanted estéril, which could be as much as fifty meters in depth.[37] This undertaking was then followed by the removal of cascalho. As Mualesso Gaston, who began at Diamang in the 1950s, stated, "Groups would come in and do jobs consecutively; one group would be in charge of removing overburden and then behind them would be a group that removed the cascalho. Overburden removal was the worst, least desirable, and hardest work. It was brutal."[38]

Prior to mechanization, laborers relied exclusively on shovels to remove both the barren overburden and underlying cascalho layers. Consequently, during Diamang's early years, the company primarily targeted creek valleys, characterized by relatively thin layers of overburden. As a result of the steady growth in the workforce, in 1928 laborers removed 603,838 cubic meters of overburden, compared with only 5,069 in 1917, the year mining operations began. The work was challenging, and the company's operational strategy was to use waves of manual laborers to perform excavation tasks.

Diamang officially entered the mechanical excavation age in 1937 with the introduction of the Ruston type 10 RB steam shovel, and added the more effective Ruston 17 RB two years later. The impact was so negligible, though, that by 1942, this machinery accounted for only 1.8 percent of the overburden removed.[39] Otherwise, virtually all of the removal was still undertaken by manual laborers. Moreover, because the (minimal) equipment that the company introduced was both unreliable and unevenly applied, many mines continued to be machine-free.

After the initial introduction of machinery, it was not until after the Second World War ended and the international demand for diamonds rebounded that Diamang sought to increase the volume of overburden and cascalho it could remove. The company's measures included further mechanization and new excavation techniques, yet it continued to rely heavily on manual laborers as it expanded its operations, thereby tempering the impact of the new machinery. As Muhetxo Sapelende, who began at Diamang in the 1940s, explained, "I was a shoveler. The company depended on our strength so there was little rest during the day."[40]

In 1947, Diamang introduced hydraulic monitors that relied on the combined power of water and gravity to blast off overburden. The monitor, or "water cannon," worked like a high-pressure fire hose, shooting out water through a handheld nozzle that had been channeled down from higher elevations.[41] African operators employed monitors when adequate gravity allowed, especially in areas where the overburden depth was greater than 4.5 meters. In 1949, company officials reckoned that a monitor could remove twice as much overburden as a team of shovelers, while using half the manpower and at less than half the cost.[42] However, only the handful of African employees who operated and maintained the monitors directly benefited; the company simply reassigned other workers, shovels in hand, to nonmechanized mines.

In the 1950s, Diamang expanded its mechanization efforts by adding scrapers, (bull)dozers, and draglines to its fleet of excavation machinery. Consequently, the percentage of overburden removed mechanically increased from 6.5 percent in 1950 to 44.3 percent in 1960.[43] Yet thousands of shovelers were still required to account for the remaining 25 percent of overburden that was removed manually during the decade. Unlike concessionary companies elsewhere in Portugal's African empire, Diamang had sufficient capital to make sizable operational investments. However, the company remained largely content to utilize these inexpensive manual laborers rather than fully mechanize its operations.

Shovelers typically worked in areas in which the overburden depth was inferior to 1.5 meters and also increasingly in new mines in western Lunda, to which Diamang would introduce machinery only after existing mines were exhausted. Luciane Kahanga, who toiled in these western stretches, said (please retain indicated), "I worked out in the newer mines shoveling . . . because the newer mines did not have all of the proper equipment."[44] However, many laborers in the company's long-standing eastern zone continued to wield shovels to complete their daily tasks as well. For example, Rodrigues, who began working on Chiluculo mine in 1968, indicated that "our job on the mine was to dig holes with a shovel. . . . We just kept digging down and down; there were many different layers and depths of cascalho. There were no machines at the mine, only shovelers."[45] Informants indicated that they were still removing both overburden and cascalho using shovels into the 1970s, in the same manner that their predecessors had for decades. Thus, although the company's production statistics reflect the mechanization of increasingly higher percentages of the excavation and transport processes, it does not follow that new equipment was correspondingly lightening workers' loads. Rather, Diamang simply redeployed most shovelers to mines at which it had not yet introduced machinery or otherwise instructed them to continue extracting what they could, no matter how minuscule the amount.

Inclement weather did little to check Diamang's unremitting push for diamonds. Beyond the first two decades of the company's operations, mine supervisors postponed production only during the heaviest of rains. Prior to the 1930s, though, when wheelbarrows were still being used to carry away overburden and cascalho, Diamang generally had little choice but to suspend operations when severe precipitation beset the region. However, as the company gradually replaced wheelbarrows with *vagonetas* (mine trams), driving rains disrupted operations less frequently.[46] Several informants indicated that they had to work during even the most relentless rains. As Mualesso Gaston recalled, "The company gave each worker a raincoat and a rain hat, so there were no excuses if the rain was hard. Sometimes we would arrive for work and the mine would be temporarily closed due to hard rains, at which point we waited until the rain passed and then started to work; we were never dismissed. . . . Often, we were made to stay and work late to make up for any time lost due to the rains."[47] Former mine workers recalled these inclement days as some of the most difficult and unpleasant they endured.

AFRICAN MEN AT WORK: TRANSPORTING THE BOUNTY

After workers stripped away and discarded the overburden, the next step entailed removing the cascalho and transporting it away to be refined at *lavarias*, or "washing stations." Methods used to complete this task ranged from carrying cascalho in woven baskets to using large capacity dump trucks. Gradual mechanization in this area of the labor process constitutes an example of technology significantly benefiting both the company and the African workforce: it saved laborers from having to transport gravel manually and also greatly increased worker productivity and, thus, operational efficiency. However, even into the 1970s, African employees laden with full baskets of cascalho on their heads trudging in the direction of washing stations remained a common sight; as always, manual labor remained the least expensive option.

For roughly the first decade of Diamang's operations, both contratados and voluntários transported cascalho in large woven baskets on their heads. This technique was extremely laborious and inefficient, and, therefore, in the mid-1920s, Diamang began introducing wheelbarrows to expedite the process. By the end of the decade, the company had also introduced vagonetas pushed along mobile "Decauville" rails, thereby greatly reducing the amount of effort necessary to transport cascalho (figure 4.1).[48] By the middle of the 1930s, Diamang had further improved this process by introducing powered cable systems, or *cabos*

FIGURE 4.1. Loading cascalho into mine trams, Miuta-1 mine, c. 1950. *Source: Science Museum of University of Coimbra.*

sem fim, that mechanically moved the trams along the rails. The inexpensive and efficient nature of this system led the company to eventually implement it across its network of mines.[49]

With each technological innovation, the transportation of cascalho grew less taxing. Consequently, Diamang increasingly doled out more of these jobs to voluntários. As always, company officials hoped to attract more local residents as voluntary workers to build a stable, proximate labor force by extending to them, when feasible, the lightest of the mining-related jobs.

The most significant introduction of machinery into this portion of the labor process was the large capacity (c. 30 tons) dump truck, which first appeared in the 1950s and coexisted with mine trams until the end of the company's operations. As with the introduction of the trams, however, the arrival of dump trucks failed to herald an end to laborers removing cascalho manually.

This manual continuity was especially visible in the new mines in western Lunda. Dinis dos Santos Muriandambo explained that when he began working in this region in 1970 he was made to carry cascalho using a basket.[50] Similarly, Luciane Kahanga stated that "after Diamang expanded to Cafunfo [in western Lunda] . . . I worked carrying cascalho on my head . . . just like in the classic [original] mines."[51] Other times, the company reinstituted manual techniques while heavy

machinery was either being repaired or undergoing routine maintenance. For example, following a 1970 visit to Capala mine, the inspector Lutero Almeida casually noted in his report that "the overhaul of
the 22 RB is done on Saturdays. In its absence, 61 men are required
to carry cascalho."[52] In response, António Ramos, a company official,
replied unsympathetically that "if we rule out the manual transport of
cascalho, the overhaul of the 22 RB becomes difficult. . . . It also breaks
down frequently, which requires us to utilize manual labor."[53] The
introduction of machinery on Diamang's mines had, once again, failed
to sound the death knell for manual procedures.

AFRICAN MEN AT WORK:
REFINING THE BOUNTY

At lavarias workers refined the cascalho to remove nondiamond elements, thereby producing a diamond-rich concentrate, or *concentrado*, that included less than 0.2 percent of the original gravel. From
the lavarias, workers then moved the concentrate to highly secured
selection stations. At these stations, "pickers" manually selected and
removed diamonds from the concentrate, at which point Diamang sent
the stones to Europe to be polished and cut.

Over time, technical innovations lightened this portion of the labor
process, though it was already composed of, by some distance, the least
taxing jobs on the mines. As such, Diamang assigned voluntários to the
majority of them, most often employing boys between the ages of ten
and twelve, whom officials deemed were less likely to pilfer stones from
the refined concentrate.[54] Yet these employees were also the most highly
scrutinized, as company officials correctly understood that exposure to
cascalho in increasingly refined forms would entice some workers to
abscond with stones. Consequently, pickers were the only Diamang
employees to be "locked down" for extended periods, isolated from the
general population and forced to remain within secured areas during
their months-long shifts.

The first task in the refinement process was to "wash" the cascalho at lavarias. The earliest and most rudimentary form of refining
cascalho was done by using jigs, which relied on water and gravity to
separate diamonds from nondiamondiferous materials.[55] Diamang next
introduced "pans," which were employed both separately and in conjunction with jigs. Whether manual or steam-powered, the cascalho fed
into the pans were subjected to a series of rotating blades that gradually

FIGURE 4.2. Loading a lavaria near Luaca mine using wheelbarrows, 1927. *Source: Science Museum of University of Coimbra.*

worked the denser material (in this case diamonds and other minerals with similar densities) to the outer edges of the pan while disposing of less dense material, such as quartz.[56] Figure 4.2 captures the manually intensive nature of lavaria work during the company's first decade of operations, highlighting the ongoing need for droves of manual laborers even after this processing technology was introduced.

The company innovated the cascalho treatment process for the last time in 1955 with the introduction of a dense-medium separation machine, or MD (from the Portuguese *meio-denso*), which relied on a Ferrosyllicate compound to separate denser materials from less dense ones.[57] These machines could treat more cascalho than pans, while also missing fewer diamonds, thereby creating a "purer" concentrate that featured a higher percentage of diamonds. By 1956, the company's solitary dense-medium separator was treating 30 percent of all cascalho, while pans continued to handle the rest.[58] It is interesting to note that, in 1974, Diamang officials were still actively pursuing plans to phase in more MDs, suggesting that they did not anticipate major interruptions in the company's operations, even though the prospect of Angolan independence, and the uncertainty it would bring, was undoubtedly conceivable at that point.[59]

The final activity in the refinement process entailed the selection of diamonds from the concentrate, which Diamang would then send out to be cut, polished, and eventually sold throughout the world. Over time, the company adopted several techniques to select and remove diamonds, but human identification and removal always remained a part of this process, which brought "pickers" into close contact with the stones. Consequently, all of the procedures associated with the concentrate featured the highest levels of security and scrutiny. As Silvestre Muachembe, who worked at a lavaria in the late 1960s and on into the 1970s, explained, "Both the lavarias and the selection stations were very restricted and highly supervised areas. . . . Treated soil [concentrate] even left lavarias in sealed carts to travel to the selection stations."[60]

Initially, workers' selections were truly rudimentary and inexact, as they were made outdoors, near excavation sites. Yet even when the pickers moved out of the sun into semipermanent structures in the mid-1920s, the process was hardly more sophisticated. Parkinson described the procedure as follows: "The pickers sat on little stools with a bench in front of them on which was placed the diamond pan and a small amount . . . of the concentrate. . . . Sometimes they used magnifying glasses. . . . Several handfuls were picked over at a time, using a small sharp stick or reed. . . . This work was done underwater which caused the diamonds to become more lustrous and to stand out more decidedly."[61] By the end of the 1920s, there were four of these selection stations in place, each one servicing a group of active mines. However, given Diamang's acute suspicion of its pickers, from the late 1920s plans were under way for a permanent, maximum security Central Selection Station (Estação Central de Escolha). Until then (and afterward), the company placed pickers under constant vigilance at work, prohibited them from having long hair or fingernails and rigorously inspected them—to the point that their clothes could be removed and burned, both when they entered and as they left. This level of security and attendant anxiety even extended beyond the walls of the selection stations. Paulo Leão Vega explained that "when you were working near or even walking by the Central Selection Station, you were not allowed to bend over and pick anything up off the ground—even your own cigarette you might have dropped—because the company would think it was a diamond!"[62] Diamang ultimately constructed this centralized facility at Andrada (south of Dundo, much closer to the mines than were the company's headquarters), and by the early 1930s it began serving as the sole selection center, with all concentrate securely transported there.

Although refinement procedures at the Andrada facility remained fairly consistent, the company's vigilance methods were more dynamic. Leonardo Chagas, Diamang's security chief from 1961 to 1974, boasted to me that he had introduced stationary gloves for the pickers, through which they could select diamonds but had no direct human contact with the stones.[63] In addition, Carlos Machado, another senior security official at Diamang, employed from 1950 until 1974, declared to me that following a company-sponsored trip to Israel to observe security measures, he was inspired to install security cameras at the Diamang selection station.[64] According to Chagas, Diamang also forced pickers to ingest laxatives and subjected them to body cavity searches.[65]

Due to pickers' unique standing within the company, their occupational calendar was radically different. Working in secured centers, they were legally locked down for up to ninety days at a time, an exception in colonial labor law that Diamang had pushed the state to accept. From virtually the inception of these centralized stations in the 1930s, three eight-hour shifts were the norm; pickers worked, slept, ate, and otherwise existed in these secured areas for months at a time. Chagas, the former Diamang security chief, indicated that as a theft-prevention measure he eventually ordered the amendment of the three-month system so that workers didn't know exactly when their stint would end.[66] Otherwise, this was a remarkably durable labor schedule that the company both honored and rarely altered.

THE ORGANIZATION OF THE WORKDAY

Over the course of Diamang's operations, shovelers worked Monday through Saturday, with a day of rest on Sunday. Mining shifts lasted roughly nine hours, starting around 7:00 a.m. and finishing up in the late afternoon. In the middle of the shift, workers typically received a timed, thirty- or sixty-minute break for lunch, which, until Diamang achieved food security in the early 1930s, was occasionally affected by food shortages. This schedule was dictated by a series of colonial decrees that were intended to maximize Africans' labor, while ostensibly protecting them against excessive work demands. Certain mine supervisors ignored these regulations, but for the most part company officials insisted on only the legally mandated number of hours from its African laborers, especially from the 1930s onward. In the 1940s, Diamang introduced the *tarefa*, or task, system on certain mines, which allowed shovelers to leave their posts as soon as they had excavated a predetermined amount of overburden and/or cascalho. This measure benefited stronger workers, while weaker ones were often forced to work longer hours and/or to enlist the help of others to avoid potential punishment.

Most shovelers rose early and arrived at work between 6:30 and 7:30 a.m. Initially, many African employees apparently struggled both to appear by their set start times and, more broadly, with the notion of industrial time, and often had to run to work to avoid arriving late.[67] To curtail this behavior, mine overseers marked anyone who arrived late in the morning as "absent," depriving them of a day's salary and, thereby, quickly ending this habit.[68] Early start times enabled workers to take advantage of the relatively cool morning hours before the sun moved overhead. In fact, some mine bosses allowed employees to arrive even earlier, which enabled them to complete more hours during this relatively agreeable time of the day. For example, when a mine inspector chided the manager of Maludi 6 mine in 1946 for encouraging this practice, the overseer defended his policy by citing elevated productivity.[69] Most laborers, however, reported to work just before an established start time, which was marked by a bell or siren. Workers then filed to their regular tasks or were assigned new ones depending on the particular needs of the mine.[70] According to Caiombo Jombe, "There was roll call each morning when we arrived. The mine boss would be given the list of who was there and then would allocate workers in groups according to who was present. So, there were many workers assigned to each task."[71]

By mid-shift, employees received a break to consume a meal that invariably featured some form of cooked manioc. Most workers received their work-site meals in cafeterias, which began to feature on company mines as early as the late 1920s, and were fixtures on most mines by the 1940s.[72] Beyond a place to consume food, cafeterias also provided shelter from the sun and, during the rainy season, from precipitation as well. Yet, at times, the construction of cafeterias (and kitchens) lagged behind the opening of new mines, and, thus, workers often began laboring at new mining installations before the erection of these facilities.

Of greater concern to workers than the existence of a cafeteria was the provision of their full, midday rations. Although the achievement of food security was a central component of the company's paternalistic push starting in the 1930s, before this time periodic food shortages threatened production. A letter from a Diamang engineer in 1924 highlights the acute concern that insufficient food supplies could generate. "We are sitting very lightly at Maludi mine and a very small disturbance in this section will close Maludi for the balance of the year. We have . . . no visible food supply. . . . Under the above conditions, you can see our situation at Maludi and how easily a false move would shut down this mine for several months. . . . If we are able to obtain the food, well and good; if not, then it will be necessary to let the workmen return to their villages."[73]

By the 1930s, moments of anxiety about the food supply virtually disappeared and even during periodic shortages, workers' lunches remained the company's highest priority. From the early part of this decade, Diamang also began diversifying meals and introducing new items, including coffee and peanuts, according to workers' tastes. As such, the nourishment of Diamang's labor force diverged significantly from the situation that Charles van Onselen described on Rhodesia's colonial-era mines: "If the death rate due to syphilis, phthisis and underground accidents was alarming, the three causes together accounted for deaths [only] in the hundreds; but the industry's parsimony over food and accommodation produced thousands of deaths. The greatest killers of all were diseases which could be directly attributed to the inadequate diet and poor standards of accommodations for workers: dysentery, inappropriate diet and scurvy."[74]

Although the provision of adequate food supplies most powerfully dictated how workers experienced their lunch break, mine managers also shaped this undertaking, especially when they imposed seemingly wanton rules such as assigning seats or prohibiting conversation in mine cafeterias. Figure 4.3 captures a somber environment in which two mine supervisors menacingly loom to ensure compliance with a "no talking" policy seemingly in effect. According to informants, these particular restrictions were not universally applied, though they were reasonably common. Janette Pedro, who was employed in the Lapaço mine kitchen from 1960 to 1961, the mine on which her husband also worked, indicated that they were unable to converse during lunch because "lunch was carried out in complete silence. No one was allowed to talk. Whites would walk through the aisles."[75] Other female informants indicated that they were simply too busy to engage in conversation. Mulevana Camachele, who also worked in a mine kitchen from 1960 to 1961, explained, "I did not visit with my husband at lunch because whites discouraged it. . . . They wanted the workers eating and the women working. There was no time for chatting."[76] After the midshift meal, laborers returned to their tasks to complete the remainder of the standard nine-hour workday.[77]

FIGURE 4.3. Wife of a contratado employee serving lunch to contracted laborers on an unidentified mine, 1962. *Source: Science Museum of University of Coimbra.*

By the 1940s, Diamang had introduced work schedules on some mines that ended when a designated task, typically the removal of a certain quantity of overburden or cascalho, was completed, rather than after a pre-established number of hours.[78] This task, or tarefa, system naturally benefited those workers who were physically capable of finishing their assigned tasks before the end of a normal nine-hour shift, while penalizing laborers who required additional time. From the company's perspective, this overhaul of the labor process ensured that workers met minimum daily targets, even if, for some African employees, it reduced the overall number of hours worked.[79]

Mine bosses and inspectors generally cooperated to ensure that tarefa targets were not excessive, though irregularities did occur, especially in cases when the cascalho was particularly dense. In 1947, for example, the mine inspector Noronha Feyo noted that on Calemba mine "we have a problem because mine overseers are setting task targets too high. . . . Many workers have to work well past 4:30 p.m. [the scheduled end of the standard nine-hour shift] to reach them, and the bosses do not deny this. Of the forty-one workers on the overburden team the day I was there, at 4:30 only four had finished their tasks."[80] Mulombe Manuel confirmed that when he began with Diamang in the 1960s, these situations still existed. "I was a shoveler of cascalho. I would dig holes with a shovel—[there were] no machines. I had to complete two square meters each day. I had a job, and if I didn't finish it I would just stay until I did. You could still be there hours after your colleagues left."[81] Much more common, however, were scenarios in which mine bosses established achievable daily assignments in the interests of both hitting monthly targets and maintaining workers' morale. Feyo's inspection of Mugigi 2 mine in 1946, in which he interviewed the mine boss, Silvino Balbino, is exemplary. According to Balbino, "I always set the target goals within the laborer's physical possibilities. When, for example, a man is fatigued—which is often a sign of illness, I make him go home. . . . I also always avoid making targets out of reach; there's nothing more demoralizing than the imposition of a goal that a worker does not have the least possibility of completing inside of regulated working hours. . . . A target that permits a worker to complete his task a little bit or even significantly before the end of the workday is imperative."[82]

RELATIONS WITH EUROPEAN AND AFRICAN MINE OVERSEERS

Both black and white mine overseers shaped the conditions under which African men and minors worked according to the varying degrees of violence they displayed. When white mine supervisors didn't administer punishment themselves, they either delegated the task to African overseers (*capitas*) or sent workers to local colonial authorities to endure it. In the early decades of Diamang's operations, both African and European overseers could generally dole out violence with impunity. However, over time, the company began punishing and then removing overseers with violent tendencies, and by the 1960s it was extremely rare for a worker to suffer physical punishment. This attitudinal shift coincided with both the company's increasing paternalism and the workers' growing sense of professionalism that, over time, would come to characterize the African labor force's approach. In practice, the latter development saw the disappearance of many of the actions that had elicited physical abuse in the first place.

Prior to these transformative developments, white overseers administered physical abuse for perceived infractions, poor performance, or even arbitrarily. For example, writing about his experience at Diamang in the 1920s, Parkinson indifferently reflected: "To strike a native in the mouth was a somewhat dangerous thing as one could cut one's knuckles very badly and might even bring on a case of poisoning. I still have a scar on my left knuckle which is a souvenir of such an occasion."[83] Not only did company officials condone corporal punishment, they also fretted over the potential loss of white employees should they pursue any punitive action against these aggressors. Occasionally, particularly violent overseers scared away current or potential African workers, but even then Diamang would typically only fine and/or transfer the transgressor; instances of outright dismissal were extremely rare.[84]

White overseers also directed violence toward the African minors on staff. For example, in 1947, a Portuguese employee, António Jacinto Cabral de Magalhães Queiroz, ordered that corporal punishment be delivered on a minor named Limão at Luxilo I mine, causing the latter to miss thirty-five days. According to the incident report:

> The African, in charge of a signal regulator for the passage of vagonetas, was found lying down near the signal . . . and the place where the mine trams passed was therefore functioning irregularly. Queiroz ordered a capita of the mine, Muamochito . . . to make Limão lay down on his stomach and to hit him on his back with a rubber tube . . . which he did; but after the second blow the minor fled and in the haste of the flight he fell on top of some planks, causing haemarthrosis [bleeding into the joints] of the right elbow, which led to sickness and an inability to work for thirty-five days.[85]

Thus, just as minors were performing useful "adult" work on the mines, they were also subject to the same labor conditions as adults, including physical abuse.

Although African employees were still enduring abuse in the 1940s, instances of it were becoming increasingly rare and European overseers were almost certain to be punished, at least financially, for any act of violence. For example, in 1947, following an incident in which a Portuguese overseer, António Emílio da Almeida, abused an African employee named Kaukuimba that resulted in the latter's hospitalization for fourteen days, Diamang fined Almeida 550 angolars and denied him his annual bonus, while continuing to pay Kaukuimba during the time he spent in the hospital and also counting this stint toward his overall contractual period.[86] Exemplary of the shifting response to African "misconduct" were interactions such as the one outlined in a report from 1946 during an inspection of Furi 1 mine, managed by a Portuguese named Calçada:

> When Calçada was sitting down for his breakfast, he was approached by two very excited workers. One of them, a strong Chokwe, already along in his years . . . complained about his companion, a young boy, who was turning over dirt with his shovel when it [the dirt] hit him as he was passing behind him. This constituted, in his estimation, a grave lack of consideration for a man of his age, and that, moreover, this was his fourth time working on the mines. Sr. Calçada listened calmly and attentively to the complaint, simply ordering that he should lower his voice. He reprimanded the novice, confirming again what he constantly recommends: the head accompanies the movement of the shovel and the throwing of the dirt. He counseled them briefly about not repeating what precipitated the complaint and then sent them back to their duties.[87]

Not only does Calçada's approach represent a marked shift in Diamang's administrative methods from earlier decades, it also highlights both the willingness of African employees to approach and register complaints with company officials and the increasingly paternalistic nature of the enterprise toward its African workforce.

It wasn't until the 1960s that physical abuse would virtually disappear from Diamang's mines—a decade during which Portugal came under increasing international pressure to curb some of its more egregious labor practices.[88] By this time, Diamang's director was characterizing the relationship between European and African employees as "a type of 'paternalism,' well understood and largely exercised, which the negro appreciates."[89] Incongruent with this approach, however, were the occasional acts of violence that appeared in company reports from this period. For example, a 1965 report outlined incidents of physical abuse carried out on Luapossa mine by a Portuguese overseer named Domingos Andrade, "who distinguished himself through his ferocity and who is a specialist in punching and especially the drawing of blood through the nose."[90] Yet the increasing paucity of such reports, coupled with oral testimony from workers indicating that abuse was either rare or nonexistent, suggests that the era of violence on Diamang's mines had largely concluded by this time. According to Joaquim Muamungo, who worked intermittently for Diamang between 1956 and 1966, "Over this period, conditions improved, and there was much less physical punishment," while Mulombe Manuel proclaimed that "there was no more physical punishment or abuse after 1964."[91]

Prior to the 1960s, laborers also suffered at the hands of African capitas, who typically came from the ranks of veteran voluntários. The violent behavior they often exhibited and their desire to avoid punishment themselves are difficult to reconcile and, thereby, render capitas extremely complex and ambiguous historical actors. Although they could be as liberal with the whip as European employees, more often capitas were often simply following orders issued by their (white) superiors, trying to avert any potential repercussions for failing to perform their duty. For most capitas, racial affinity superseded any existing or potential ethnic animosities, and, thus, black overseers regularly identified with the miners they oversaw, even as they administered corporal abuse. Dinis dos Santos Muriandambo, a former capita, recalled: "It was difficult to be a capita because I had to physically discipline comrades. At times, I would pretend that I didn't know those who were . . . doing something punishable."[92]

In general, workers tolerated capitas who dealt with them fairly, even if this treatment at times included corporal punishment. Caiombo Jombe, who was at Diamang intermittently from 1964 to 1975, articulated this type of understanding, declaring that "capitas would hit workers, but really only with sticks [read: not whips] on the shoulder or back to urge them to work faster."[93] For Domingos Matos, the relationship was simple: "The capita would not get mad if you did not do anything wrong."[94]

Ultimately, African capitas shaped the working environment for laborers largely according to their individual propensity for violence, much as white bosses demonstrated varying degrees of aggression to promote discipline and improve production. Testimony from Muatxinjango Maca, who worked at Diamang in the late 1940s, captures this range of personalities and experiential diversity well. "If you did not finish work, something bad would happen to you. . . . You would be taken . . . to be beaten by the chefe do posto. There, you were made to lay face first on the ground and be beaten with a *chicote* [whip] or stick. . . . If you had a "good" capita then he would take you to the posto, but if you had a "bad" capita he would begin the beating on the spot, and then *still* take you."[95]

MINERS AT WORK: OCCUPATIONAL INJURIES

Work-related injuries constituted a serious, even lethal, risk for African laborers, most of whom daily participated in the most dangerous of the core mining processes: the excavation and transport of gravel. And because more contratados than voluntários performed these tasks, the former also suffered a disproportionate number of injuries. Yet because shovels, baskets, and wheelbarrows were much less perilous than mine trams and heavy equipment, the sluggish nature of mechanization on Diamang's mines actually served to reduce the overall quantity and severity of injuries. When workers did incur injuries, Diamang sought to contain both the attendant short- and long-term costs, including treatment and compensatory payments, respectively, for which it was legally responsible. In practice, Diamang's mine administrators were more focused on the profits that workers helped produce rather than on the problems, such as injuries, that the latter suffered as a result. As Parkinson reflected, "In the 1920s, our primary purpose . . . was quite single-minded: we were supposed to find and produce diamonds. This left little time for worrying about the welfare of the natives except in so far as their efficiency was concerned."[96] Yet for all of its attention on profits, even before the 1930s (and especially afterward), Diamang rarely failed to honor its accident-related responsibilities. Unlike on mines elsewhere in Southern Africa, Diamang dependably treated injured workers, providing both immediate compensation and monthly pensions for those employees who suffered debilitating injuries.[97]

The Angolan company's construction of lavarias produced the first spike in workers' injury rates. A 1929 incident in which Salonga, an African employee working at the lavaria on N'Zargi III mine, fell into the basin of a pan and had his left foot crushed by the rotating knives was typical of the type of injury that occurred at these sites.[98] Later that same year, two similar incidents occurred: an employee named Chumanga lost his right hand, while another, Mussumari, fractured his.[99]

Into the 1930s, the company's introduction of mine trams, especially those with powered cable systems, generated an increasing number of injuries, raising the annual number of reported injuries from 52 in 1934 to 130 in 1937, including 16 deaths. In 1948, Diamang produced a comprehensive study detailing the previous six years of mine injuries, which further implicated trams as the leading danger to mine workers (table 4.1). According to many doctors and mine bosses, mine tram incidents were often caused by workers' "lack of care."[100] Informants indicated, however,

TABLE 4.1

Work accidents that led to death or more than fifteen days of hospitalization

Cause	Number	Percentage
Mine trams	371	44.48%
Miscellaneous	229	27.46
Falls	61	7.31
Machines and gears	58	6.95
Shovels and other tools	42	5.04
Traction cables	35	4.20
Transmission belts	20	2.40
Collapses	11	1.32
Transportation	5	0.60
Electrocution	2	0.24
Total	834	100

Source: Diamang, *Determinação de um indice capaz de traduzir por empregado o seu interesse pela questão acidentes de trabalho* (1948), 1, MAUC, Folder 86 52°.

that trams were heavy and difficult to push, especially uphill, and consequently they often tipped over. Figure 4.4 captures the challenging nature of this work. As Deque, a former tram operator in the 1950s, asserted, "When these types of accidents occurred, people could, and did, lose legs or were even cut in two."[101] João Paulo Sueno's experience perhaps best highlights how the company created scenarios in which trams could overwhelm, and thus injure, their operators. "I was eight years old when I started to work with the company [in 1963]. I began by working as a servant in a house . . . and learned Portuguese. . . . Later, I went to Cula mine and worked with mine trams. I was eleven when this happened! . . . I admit, it was difficult because I was younger. . . . For four years I worked with trams."[102]

Due to the rapid expansion of operations and the corresponding increase in the number of inexperienced workers, in 1945 Diamang reported 164 work accidents. It is important to note, however, that these figures included only "serious" injuries as, for example, a November 20, 1946, inspection of Caúma mine alone revealed 57 workers with injuries, out of a total of only 209! Both the inspector and a company doctor characterized these injuries as "small," implying that Diamang required employees to work through them.[103] By 1972, mine-related injuries had reached 326 annually (totaling 14,678 workdays lost, not including deaths or permanent incapacitations),

FIGURE 4.4. Young tram operator on Cassiaxima-3 mine, 1946. *Source: Science Museum of University of Coimbra.*

though the number of unreported, or "small," injuries was surely greater than this figure.

After an injury was sustained, the company required mine overseers to determine the victim's level of culpability (if any) so that it could calculate a compensatory amount, or "pension," which was mandated by the colonial Native Labor Code, or CTI (Código do Trabalho dos Indígenas). Company representatives first had to ascertain the cause of the accident in an attempt to establish who, or what, was to blame.[104] For example, concerning the injury to Salonga cited above, company doctor Vasco José de Oliveira determined that "because Salonga did not injure himself intentionally in order to remove himself from work, was not drunk, and did not forget the recommendations made concerning how to perform his job, he has a right to a pension."[105] Conversely, de Oliveira recommended that in Chumanga's case (also cited above), he had "carried out his job with so little skill that the accident had resulted, and because . . . the victim himself declared that he was warned daily by the engineer of the mine about the misfortunes that could befall him if he did not carry out his job with a great deal of attention . . . no compensation is owed to him."[106] Although it is unclear if Chumanga ever received any payments, records indicate that as of December 1929, Salonga began receiving a life-long, monthly pension of 25 angolars.

As injury rates rose, so too did payouts. In 1934, for example, Diamang distributed 8,460 angolars in injury-related pensions, while this figure jumped to 18,195 in 1939. Alarmed by these increasing costs, in 1941 the company attempted to minimize what it was paying out by introducing a grisly chart that determined compensatory sums according to what body part had been damaged or lost and how occupationally important that appendage had been. For example, the loss of a finger warranted a pension of only 5 percent of a manual laborer's salary, but 20 percent if the employee was a clerk. Diamang valued whole hands more, paying out 45–60 percent according to occupation, though it capped all compensatory payments at 60 percent.

Into the 1940s, the company began compensating workers with lump sums, either in lieu of or in conjunction with monthly, lifelong pension payments. In line with these new policies, Diamang paid an African employee named Luiz, who had lost his right pinky finger on Mugigi mine in 1949 in an accident, 720 angolars up front and then a finite monthly pension of 4 angolars. By the 1960s, Diamang was respecting a 1956 update to the CTI, which dictated that companies determining monthly pension amounts must incorporate one-third of a worker's salary; take into consideration both the degree of incapacity and the age of the victim; and provide an up-front compensatory payment of three full years of the injured worker's salary. However, informants indicated that the company still occasionally shirked its responsibilities. For example, Mário Alfredo Samuhaniquime indicated that "an eye injury I incurred while using a prospecting drill in 1973 ended my career. The company gave me no compensation. The day I walked out of the hospital I was done receiving money."[107] Samuhanquime's testimony suggests that company officials' desire to minimize operational costs occasionally caused them not only to ignore colonial legislation but also to lose sight of their own overarching labor attraction and retention objectives.

For those workers who avoided severe accidents but still required medical attention, the point of access to Diamang's health services was the *posto de socorro*, or first aid post, which several mines featured. These posts were manned by African health auxiliaries who were capable of treating basic injuries, namely simple wounds, but who would pass along more difficult cases to regional clinics or hospitals staffed by more capable nurses and, ultimately, Portuguese doctors.[108] The interaction depicted in figure 4.5 was most likely staged, though the treatment being administered is consistent with the type of assistance that

FIGURE 4.5. Treatment at a posto de socorro, 1938. *Source: Science Museum of University of Coimbra.*

these auxiliaries provided. The company introduced these stations in the 1930s, though never uniformly, and some informants indicated that the mines on which they were working in the 1970s still lacked these facilities. Furthermore, even when postos did exist, accessing them was not always possible. Felipe Leo Muatxissupa, who began at Poné mine in 1966, explained: "There was a medical post on the mine; people would go there to receive treatments for accidents. . . . You would also go there for medical treatment. It was somewhat difficult to visit the post, though, because the company generally discouraged it. You had to have a serious injury."[109] Although most informants reported no such impediments to visiting these stations, one former employee, Costa Chicungo, who began at N'Zargi mine in 1952, indicated that "if you were a good worker, the company would treat you medically, no problem, but if you were a 'bad worker' or 'bad man,' then the company would report you to the government and then the government may even go so far as to actually punish you—far from helping you. Or, the company may just choose not to help you, but without (administering) actual punishment."[110]

On most mines, from the 1930s onward, monthly inspections by Diamang medical personnel were the norm and typically occurred during lunchtime, after work, or on Saturdays. These exams enabled the company to monitor employees' overall health, address any "small" injuries that had not required immediate attention, check the advance of any communicable diseases by isolating sick workers, and order food supplements for workers who were failing to maintain or gain weight, or even to have them transferred so that they could perform less taxing tasks. According to Alberto Rossa, who started with Diamang in 1949, "Once per month, a doctor would perform a general inspection of the workers. If a worker only had a light fever, for example, he would continue to work, but would be given lighter tasks such as cutting grass or cleaning. . . . A doctor could authorize certain workers to receive nourishment supplements, but these still had to be composed of standard food."[111] While these examinations were not uniform, most informants reported that they were part of the routine of mine life and, thus, they can be understood as another effort by the company to safeguard its investment in the African labor force.

MAXIMIZING MINIMAL COMPENSATION

In 1949, the governor of Malange Province, an area immediately west of Diamang's operational zone from which the company regularly drew contratados, described Diamang's wages as "the lowest in the entire colony."[112] Although not technically correct, the accusation does reflect the contempt that some colonial administrators felt toward the extremely profitable—and powerful—diamond enterprise and is, anyway, reasonable in the sense that Diamang was offering wages that were definitely lower than it could afford to pay. Forever rebutting these accusations, company officials were always quick to point to the wide range of services it offered its African employees, which were unparalleled in Portugal's colonial empire. Moreover, the tax rate in Lunda was among the lowest in the colony, as was the cost of living due to the suppressed prices at Diamang's line of stores and the fact that, at any given moment, the company was feeding, housing, clothing, and otherwise caring for a significant number of the region's residents.

This comprehensive strategy instantiated the enterprise's paternalistic approach to its African employees and, more broadly, to the entire population of Lunda. Given that local residents lacked other employment options to meet their tax requirements, Diamang understood that a minimalist approach to salaries, augmented by substantial payments in kind, was sufficient to maintain regional stability. As Peter Carstens has contended in regards to mining operations elsewhere in colonial-era Africa, "A very important factor contributing to worker loyalty . . . is the provision of social services exclusively for company workers— services not available outside the firm's domain."[113]

Following the launch of operations in 1917, the fledgling enterprise was offering little in the way of social services, however, and wages that were so low that a worker's entire salary could be consumed by his tax commitments. In 1925, for example, one empathetic company official reported that of the seventy-one laborers he recently received, "forty had not paid their taxes and, thus, are not paid to work, but are instead working off their outstanding tax requirements. Each day, after work, some of these laborers arrive at my house in order to ask how many days they have to remain in service. These are employees who don't have the courage to flee. But it isn't very hard to determine the cause of those who do flee: neither white nor black likes to work for free."[114]

As revenues began to rise, Diamang periodically increased salaries and, in particular, in conjunction with the broader improvements of the 1930s. From 1935 to 1937, salaries almost doubled, though going forward dramatic boosts in pay such as this one were rare. However, as long as conditions continued to improve on the mines and a favorable gap existed between wage and tax levels, worker discontent over remuneration remained minimal.

In the 1960s, partially in response to accusations of worker exploitation that Angolan nationalist groups were directing at the colonial regime, salaries at Diamang escalated both significantly and rapidly, doubling from the beginning to the middle of the decade, while the company also introduced annual raises and paid time off. Consistent with its paternalistic approach, Diamang also reduced the prices of Africans' "favorite items" at its stores. Following the outbreak of the Angolan revolution in 1961, de Vilhena alerted Salazar that Diamang was "lowering the prices of the goods, foodstuffs and other items Africans appreciate, from bowls made of enameled iron to bicycles, and, for women, sewing machines and *panos* [fabrics worn as garments]."[115]

African laborers who worked on the mines during this period remembered their compensation and corresponding purchasing power divergently. According to Joaquim Ezaia, "Even when I got my big raise [following his promotion from manual mine laborer to mechanic], there was still not much to do but eat and survive—I didn't use it to chase women!"[116] More common, though, was testimony that reflected both the paternalistic nature of the company's compensation strategy and employees' general contentment with Diamang's approach. Joaquim Trinidade, for example, declared that "the company paid poorly, but the benefits and living conditions were great, for example, health care. . . . We always had electricity, water, and housing, care for pregnant women, and many other things."[117]

GENDERING THE LABOR PROCESS: THE CRITICAL BUT OFTEN INVISIBLE WORK OF WOMEN

Although women never participated directly in the mining process, wives of contratados served as kitchen and cafeteria staff, mine encampment cleaners, and, most importantly, as agricultural laborers on the farms and plantations that fed Diamang's ever-expanding workforce. Their contributions were both critical to the social reproduction of their families and vital to the company's operations. Because Diamang did not consider the women engaged in these tasks as formal laborers, though, women's official employment figures are deceptively small— between 2 and 5 percent.[118] Yet, these female employees were subjected to many of the same challenges that men faced on the mines, including onerous working conditions and occasional violence, while also enduring gender-specific challenges such as unwelcome advances and sexual abuse. Without this prodigious, yet officially "invisible," labor force, Diamang would have had to divert valuable male laborers to complete a range of operationally crucial tasks.

A COVETED PLACE IN THE KITCHEN

Concomitant with the company's introduction of cafeterias on the mines was its employment of the strongest and healthiest wives of contratados in mine kitchens to prepare, serve, and clean up after midday meals. These positions were attractive to women who, starting in 1925, received a full salary and rations for their efforts.[119] Conversely, toiling on the company's plantations or cleaning encampments constituted much more laborious tasks and garnered women only rations as compensation (though Diamang did introduce a salary for plantation workers in the 1960s). Testimony by Janette Pedro, who began working in a mine kitchen in 1960, highlights the discrepancy between cafeteria work and other endeavors, as well as women's general sentiments toward them.

"My husband and I went to Lapaço mine. I first went to the kitchen to work as a cleaner, but then I became a cook. I worked on the same mine as my husband. The best, most desirable, work for women was in the kitchen, even though we had many responsibilities each day—for example, cooking, washing, etc. It was the lightest work—versus working in the fields or cleaning the village."[120] Illustrative of the professionalism that these women espoused, Mulevana Camachele stated: "At work, I would cook for the contratados. I was mainly a dishwasher but I would also cook, and of course, I already knew how to do it; there was no training required. I earned the same as my husband did shoveling, but this equal salary did not cause any problems with my husband. Couples are couples. We were there to work together."[121]

To prepare and serve mid-shift meals, women arrived at work in the morning and then cleaned up afterward; the entire process lasted roughly five to seven hours. Lina Machamba, who worked in a mine kitchen from 1967, described the workday as follows: "I served in the kitchen as a cook. . . . I went in at 7:00 a.m. and had to make a large fire with firewood and heat up a giant pot of water. Afterwards, I cooked the *funge* and then served and cleaned up after the workers. . . . It was very difficult [work]."[122]

Kitchens were adjacent to cafeterias, and each typically supported a single mine, though in some cases they generated meals for multiple mines. Staff sizes, including either a Portuguese male or African female manager, ranged from five to fifteen, generally according to the size of the mine population. Machamba explained, however, that on her second stint with Diamang her kitchen at Chilumbuka mine served approximately three hundred laborers, but featured a staff of only five people, so "life was hard."[123] By the 1940s, few open-air kitchens remained, and in the 1960s Diamang had begun to install modern machinery in some kitchens, including large ovens and dishwashers. Given the provision of both rations and a salary for this work, its relative ease, and a familiarity with the constituent functions of preparing, cooking, serving, and cleaning, women coveted these positions.

HARVESTING CHALLENGES: WOMEN AS FIELD LABORS

From the company's inception, Diamang had encouraged wives of contratados to contribute to the enterprise's food supply. For these women, this involvement initially meant cultivating personal plots near mine encampments and using the yields to feed themselves and their families.[124] However, this prescription was impracticable. Lina

Machamba explained. "I had no time to maintain a small garden plot of my own; I worked all day and then had to cook for my husband."[125] Into the early 1930s, Diamang established a series of weekly markets to which regional residents brought agricultural surpluses and received what informants uniformly described as "fair prices." In 1937, in a complementary measure, Diamang again turned its attention to the women already on the mines and newly thrust these wives to the fore-front of the enterprise's food production campaign by mandating that they work each weekday morning on company-operated plantations (figure 4.6). With this measure, Diamang acknowledged that it both

FIGURE 4.6. Female field laborer carrying manioc on a "native" plantation, 1964. Source: Science Museum of University of Coimbra.

needed the labor that these women could provide and required more contratados' wives to accompany their husbands to the mines to further increase food production. Diamang's technical manager, H. J. Quirino da Fonseca, assessed the progress of the program following its inaugural year: "By the end of the year [1937], during which it was necessary to establish among the women the habit of working regularly and hence overcome their natural reluctance, 42 hectares were planted. This year, from January to September, 250 hectares have been planted with the hope of (planting) 300 more. This means that in the first year these women . . . will have planted over 500 hectares."[126] Apparently, this "habit" was sufficiently instilled in this female workforce going forward.

These female field laborers were the strongest women available after the company had selected individuals for kitchen duty. In 1952, for example, 1,432 contratado wives out of a total of 2,232 were assigned to field labor.[127] The crops grown on "native plantations," including manioc, beans, and sweet potatoes, were for the exclusive consumption of these female laborers, their husbands, and children, and thus they had every reason to tend to them assiduously. Meanwhile, on "company" plantations, women assisted in the cultivation of several staples, including rice, vegetables, and fruits grown in orchards, though members of the European staff were the intended recipients of these harvests.

Compared with kitchen workers, female fieldworkers encountered different sets of company policies and associated daily challenges. Although agricultural laborers worked from only 7:00 a.m. to roughly 11:00 a.m. or noon before returning to encampments, fieldwork was much more demanding than kitchen work. As Sacabela Sacahiavo, whose wife worked on Diamang's plantations in the early 1950s, indicated: "Kitchen work was much more desirable because it was . . . not out in the bush where women would get cuts all over their arms because of the grass and it was brutal because of the sun and heat."[128] The company also required field laborers to strap any accompanying small children to their backs while they worked. According to Mawassa Mwaninga, who was at Diamang from 1964 to 1965, "I worked in the fields . . . from 7:00 a.m. to noon . . . with a child on my back, and then went back to the encampment."[129] Further differentiating field labor from kitchen labor was the fact that the core tasks of planting, maintenance, and harvesting were undertaken according to crop cycles, unlike kitchen labor, which was more routine. For example, laboring in the fields was more or less demanding depending on the particular moment in the growth cycle, with women taking advantage of the rainy season to work the soil. Agricultural laborers also disliked it when Diamang transferred their husbands across mines, as this relocation took them away from the fields into which they had poured so much hard work. Some company officials even regretted that these women, who might only stay at Diamang for a year, would perform much of the hard labor "but then might not really enjoy the yield at the end."[130]

A NEGLECTED CREW: CLEANING THE ENCAMPMENTS

Company officials assigned those contratados' wives whom they had passed over for work in either the mine kitchens or the fields to clean the mine encampments. This group of female laborers was composed of women perceived by Diamang officials to be the weakest, but also of wives who had children who were too old to be carried on their backs while they worked in the fields but too young to work in that capacity themselves. According to Janette Pedro, "The Portuguese selected the strongest women to work in the kitchen; those a bit older or a bit weaker were chosen to work in the fields, and then the weakest and oldest [women] and the children stayed in the village to clean it."[131] As one official put it in 1945, this group consisted of "mothers, and old or meager women."[132]

Although in the company's early years accompanying wives were responsible for cleaning mine encampments, after Diamang had created kitchen and agricultural jobs, only those women assigned to this task remained in the company settlements each morning. In general, Diamang strove to achieve a ratio of two women for every nine houses in a given encampment, but since housing figures were imprecise, it is unclear how well this intention reflected reality. In 1952, 473 of 1,794 wives, or 26.4 percent, worked in this capacity, and in 1954 the figures were similar: 592 and 2,232 (26.5 percent), respectively.[133] In many respects, cleaning mine encampments constituted the easiest of the tasks that women were assigned, but because the company never remunerated these women for their efforts this work was also the least desirable.

ABUSE OF FEMALE EMPLOYEES

While women were able to avoid the injuries that daily exposure to the company's excavation sites generated, they experienced the same regimented environment and (although to a lesser extent) physical abuse that male laborers endured. Furthermore, many women working as agricultural laborers were also forced to contend with unwanted sexual advances at the hands of aggressive European or African overseers as they toiled on isolated plantations and farms. Over time, company officials increasingly punished transgressors, significantly reducing the frequency of this form of abuse, while also privately fretting that these incidents might drive down spousal accompaniment rates.

Although it was somewhat rare, female laborers did suffer physical abuse while at Diamang. In 1959, for example, a Portuguese overseer, A. Lopes, who had a tarnished record, kicked and broke the arm of a female worker named Namutondo while she was working on a company farm. Afterward, the mine administrator, J. Robalo, filed the following report: "He claims she was working slowly; he grabbed her arm to speed her up and that she was fine. . . . Five of her co-workers, and thus witnesses, claimed that he kicked and subsequently broke her arm after she said she was old and could not work any faster, and that after she was obviously in pain he examined the arm and told her that nothing was wrong."[134] In this instance, Diamang took swift action against the problematic employee, though Lopes's alleged "good" qualities served to lighten his punishment. Mário Correira, the company official responsible for handling the case, rationalized his decision as follows:

> The qualities of A. Lopes's work are well known. The villages and farms under his charge are always well presented, and the yields from them are quite good. . . . Given these factors, and seeing that: (1) He is a very good worker, (2) He was recently punished, with a partial loss of the annual bonus, and (3) He will pay 400$00 [escudos] as compensation to the woman, or will be prosecuted by the authorities, we propose that he . . . pay all of the expenses of the hospitalization of the *indígena* Namutondo and that this incident be considered on the occasion of the next annual bonus, as well as his behavior until then.[135]

Many female field laborers also had to contend with the predations of African and European overseers, unlike most kitchen and mine encampment cleaning staff. Although few women experienced unsolicited advances, and most of my female informants reported cordial relations with overseers, incidents of this nature featured on Diamang's installations from the moment women started accompanying their contracted husbands.

In 1950, a series of incidents of sexual aggression against women generated the first major scandal of this type, prompting even de Vilhena to address this heretofore "undiscussed" issue. In this case, contratado laborers from Songo had alerted a chefe do posto that African capitas and European overseers were sexually abusing female field laborers with impunity.[136] A flurry of correspondence proceeded between the directors in the field and de Vilhena that acknowledged the severity of these perpetrations. With an eye toward the tenuous nature of the labor supply, however, the constituent letters and telegrams focused most squarely on the overall impact that these actions might have on spousal accompaniment. For example, J. Tavares Paulo, the director general in Lunda, declared that this "grave conduct by certain individuals, speaking even of European employees, constitutes the greatest of the impediments in the fight in which we find ourselves to get the workers to have their wives accompany them."[137] Another letter from Paulo dated the same day reveals a callousness, or perhaps simply a dose of skepticism mixed with realism, toward the allegations. "Of course, some of this is going to happen given the sheer volume of people involved, and as much as we would like it to be a utopia, it is not. Also . . . [colonial] administrators might be exaggerating the volume of complaints."[138] It appears that Paulo's dismissive sentiments prevailed, as the company never doled out any specific punishments stemming from these revelations.

Fortunately for female employees, very few of their interactions with African or European overseers were hostile or aggressive in nature. While company records are largely silent in chronicling amiable day-to-day relations between male overseers and female laborers, many informants spoke fondly of these relationships. As Janette Pedro, who first arrived in Diamang in 1960, explained, "I had no problems with men on the mine or in the encampment. My capita was smart, very religious, and did not trust women! So, conversation with him was minimal, and he never made any advances towards us."[139] Similarly, Anna Maria dos Santos, who worked in the agricultural division from 1974 to '75, recalled that "my boss was Portuguese, but I was treated as an equal . . . and he made no sexual advances toward me. . . . Actually, women were often 'pardoned' for things that men could not get away with; men had so little luck!" Regarding these male co-workers, she added: "I always had to prove my worth because males were showing off how strong and tough they were, so I had to, as well. I received the same salary as the men. They would protest that this was not right, and they would say that it was not fair that I was receiving the same salary as them! . . . [But] we wouldn't complain about each other because we didn't want to attract the attention of the boss. We had love for one another, and tried to work any problem out amongst ourselves."[140] Dos Santos's sentiments exemplify the pervasive professionalism that over time African employees both cultivated and exercised on the Angolan mines.

⮌

Over the course of Diamang's operations, hundreds of thousands of African employees toiled on the company's mines and in closely related supporting endeavors. The majority of laborers were men, but over time women and even minors became increasingly vital members of the workforce. Diamang assigned these laborers to different tasks, at times according to whether they were voluntários or contratados, which, in turn, shaped these individual workers' experiences. Regardless, all employees were subject to many of the same challenges, including arduous workdays exacerbated by inclement weather or stifling heat. Over time, gradual mechanization and periodic amendments to operational procedures, such as the institution of the task system mitigated the labor process. In particular, the introduction of certain equipment, such as mine trams and dump trucks, greatly alleviated specific aspects of the broader labor process, though Diamang never relented in its quest for manual laborers to work with, or simply alongside, this machinery. Mechanization also increased the number of occupational injuries that workers sustained.

Unlike the paternalistic improvements that Diamang made away from the mines, the company's ongoing insistence on predominantly manual labor processes rendered overall work-site conditions more static, even after the revolutionary events of the early 1960s. In response, many African employees creatively engaged with or, especially during the first two decades of Diamang's operations, even abandoned, the labor regime. Yet as time passed, most workers demonstrated increasingly high levels of occupational professionalism, as manifested in their resourceful, strategic, and committed approaches to their daily tasks. In the following chapter, I explore the gendered strategies that African men and women employed at their respective work sites.

5 ◐ Negotiating Stability
Laborers' Work-Site Strategies, 1922–75

> Portugal has always followed a colonial policy particularly her own, which
> has enabled it to avoid the troubles which other colonies have suffered,
> such as strikes, excessive and disorderly syndicalism, native uprisings, etc.
>
> — *Diamang engineer W. A. Odgers to*
> *Ernesto de Vilhena, director, 1947*

ALTHOUGH THE SENSATIONAL "STRIKES, excessive and disorderly
syndicalism, and 'native' uprisings" were, indeed, absent from laborers'
strategic repertoire on Diamang's mines, the quietism in Lunda prevailed in spite of "Portugal's *particular* colonial policy." Although Diamang's paternalistic initiatives and Lunda's isolation both played key
roles, African employees' professionalism, as exhibited in a wide array
of gendered work-site strategies, best explains the absence of unrest
on Diamang's mines. Male and female laborers regularly engaged in
low-risk activities, such as sharing tasks, singing songs while toiling,
voicing complaints, and redistributing mine meals, which enabled
them to cope with the daily work regimen by attenuating its harshness,
while never threatening to undermine or subvert it. A smaller number of employees chose to participate in more precarious undertakings,
including feigning illness, partially withholding their labor, and even
diamond theft, which signaled a reluctance to comply fully with the
labor regime. Still others elected to forsake the company altogether
through absenteeism or desertion, though the transformative changes
on the Angolan mines beginning in the 1930s saw the virtual abandonment of these more extreme measures.

In this chapter, I examine the gamut of strategies that African laborers pursued over time at Diamang's work sites. In the first two decades of the company's operations, employees engaged in an array of aggressive activities, either independently or in small groups. As time elapsed, however, and Diamang began introducing its range of service upgrades, while also expanding its monitoring and supervisory staffs, both the space in which to maneuver and the motivation to do so drastically diminished. Meanwhile, workers were correspondingly adopting a more professional approach to employment, elevating productivity and largely disposing of more drastic endeavors that could jeopardize their primary objective of uneventfully completing their contracts.

POPULAR STRATEGIES: MINIMAL RISKS FOR REASONABLE RETURNS

The majority of male and female workers who sought to improve their daily labor experiences openly engaged in work-site strategies such as sharing tasks, singing, redistributing mine meals, or issuing complaints about particularly disagreeable work conditions. None of these pursuits explicitly violated company policies, thereby rendering them no- or low-risk in nature. In fact, only the issuance of complaints could elicit punishment, though more often they prompted procedural changes or simply met with indifference. Both before and after the 1930s, these strategies generally served to improve employees' work experiences and were, thus, both popular and durable options.

DISTRIBUTING THE LOAD: TASK SHARING

African employees engaged in task sharing largely as a creative response to the implementation of the *tarefa* (task) system. Company officials rarely commented on this strategy in the written record, suggesting that they neither concerned themselves with how workers reached task quotas nor considered it an objectionable practice. Although Diamang officials may have been apathetic toward this strategy, informants' testimony reveals that male and female workers commonly and energetically shared tasks, which took on different forms depending on the particular assignment in the labor process.

One type of task sharing in which workers engaged was a type of "subcontracting," in which workers enlisted others to help and compensated them for their efforts. Alberto Rossa, a former overseer who regularly observed this arrangement, indicated that "those who could not finish [their tasks] would actually pay stronger workers to help them finish."[1] This was, in practice, a prudent decision, as mine overseers otherwise compelled laborers to continue working past the end of the standard nine-hour shift to complete any unfinished tasks and/or occasionally resorted to physical abuse to "motivate" these workers to finish their assignments.

To help one another avoid punishment, though, most workers readily shared tasks without demanding any type of compensation. As António Batista, a former employee, indicated, "I was required to dig two meters of overburden each day . . . and I could go home as soon as it was finished. But, when a person finished early, he would definitely stay and help others finish."[2] Sacabela Sacahiavo's testimony regarding task sharing in the 1950s suggests that Diamang played a role in facilitating this arrangement. "I worked as a shoveler digging *cascalho* [gravel]. I was immediately thrust in. In my case, though, the company alternated a new person with a more experienced person so that you always had someone next to you to help, if necessary."[3]

Irrespective of the intentionality of this work-site configuration, Diamang benefited from this spirit of cooperation among its African labor force, as even the threat of corporal punishment could not propel frail workers to remove more gravel than they were physically capable of shoveling. In practice, the high level of collaboration that workers displayed in pursuing this occupational strategy is rather remarkable considering that Diamang regularly transferred employees both across and within mines. As Mulombe Manuel explained, "We would typically help each other finish . . . but because you worked next to different people almost every day, it could be hard to make friends."[4]

Female employees also strategically shared tasks. In the fields, women willingly formed teams to better tackle the various tasks at hand. In 1962, for example, a company report indicated that "in the majority of cases, women field laborers prefer to arrange themselves in small groups of between five and ten according to their relations, affections, or connections."[5] In the kitchens, too, women took the initiative to pool their labor, even though they were typically assigned individual tasks. Janette Pedro indicated that on Lapaço mine, where she worked in the 1960s, "The tasks in the kitchen were divided equally: four cooking, four attending to other things, but periodically we voluntarily rotated and shared these jobs."[6] Mulevana Camachele, who was with Diamang from 1960 to 1961, also touched on this flexibility and cooperativeness in her testimony. "At work, I would make *funge* for the contratados. I was mainly a dishwasher but I would also help out cooking. . . . Six other Angolan women worked in the kitchen."[7] As in cases where men shared tasks, company officials were largely indifferent to this strategy, so long as tasks were completed.

Both male and female Diamang employees sang while they worked, which served to animate them or even disparage the Portuguese; in either case, they helped workers countenance their difficult days. Although informants who had formerly served in a variety of different occupations mentioned this activity, those who had performed tasks that lent themselves to the songs' rhythms, such as shoveling, cited it most often. Most Portuguese overseers tolerated singing because it appeared to improve morale and productivity. However, African capitas occasionally forced laborers to cease when songs denigrated the Portuguese, even though the lyrics were undecipherable to the European mine bosses.

Diamang officials generally tolerated, or even encouraged, singing, deeming that the morale generated by this seemingly innocuous activity outweighed any potential drawbacks. As Lute J. Parkinson explained, in the 1920s and 1930s "singing was the usual accompaniment to any strenuous job. . . . One man would sing the *obligato*, then everybody would shout in chorus . . . exerting their strength in unison. . . . To keep the men happy and their spirits high, the European rarely objected to this."[8] In fact, Bartolomeu Lubano indicated that in the late 1960s at Kandala, an active prospecting site in western Lunda, "The company employed an Angolan to beat an African drum on the work site in order to animate the workers and keep them moving and in rhythm while they sang apace."[9]

Informants confirmed that they most often sang to animate and cheer themselves up while working. According to Costa Chicungo, who began with Diamang in the 1950s, "Songs were sung mainly by contratados on the mines in order to remember family, etc."[10] In this way, singing connected workers to distant homelands, facilitating a mental escape from the challenging physical conditions in which they were immersed.

Other times, songs had a pejorative purpose. Paulo Leão Vega and António Sulessa, former employees, explained that "we definitely sang derogatory songs in Chokwe about the Portuguese [colonial administrators]," one verse of which went as follows: "The white makes us suffer, punishes us, hits us, offends us, and beats us with the *palmatória* and the *chicote*."[11] Capitas, who often spoke Chokwe, were almost always complicit, yet on occasion they would order laborers to proceed silently, further underscoring these overseers' conflicted and ambiguous position at the company. João Muacasso indicated that on Cassanguidi mine in the late 1950s, "You could not sing insulting songs against the Portuguese because of the capitas. Those who did anyway . . . received lashes with a *cahenge*, which is like a whip."[12] On rare occasions, even when songs were not intended to disparage, Portuguese mine bosses' unfamiliarity with workers' tongues caused them to be suspicious, prompting them to order the singing to halt. As Costa Chicungo explained, "Songs were not really directed at Diamang's Portuguese bosses. Nonetheless, they would get mad because they could not understand local languages and would often order workers to stop. If we continued, we would be punished, but few, if any, ever did after being told to stop."[13]

Women also sang as they labored, and for the same reasons. Mawassa Mwaninga hummed the following for me (repeated twice) as she recalled how she and her friends would sing to both cheer themselves up and mitigate the difficult days working in the fields during the mid-1960s: "I set off to sell what was once dear; the suffering takes us. Your friend who does not work; what will he eat?"[14] Again, rhythmic tasks, such as field labor, were more likely to provoke singing, but it appears that unlike company overseers' handling of male workers, they never silenced these female voices.

FEEDING THE FAMILY: REDISTRIBUTING MINE MEALS

At mine cafeterias, some workers carefully reserved a portion of their own mid-shift meal so that it could later be sold, traded, or simply redistributed to family members. To carry out this strategy, laborers typically had to tote these portions back to mine encampments after work. For a brief period in the 1940s, however, certain mine bosses distributed ration entitlements directly to wives of contratados who assembled at mine cafeterias each day, and thus these women could also carry away any surplus food that their husbands had set aside. Over time, mine overseers largely tolerated this activity; conversely, most mine inspectors, as well as company medical personnel, disapproved, fearing that laborers were depriving themselves of much-needed nourishment. By the 1950s, Diamang's increased ration allotments for family members obviated the pursuit of this strategy and concomitantly ended the intra-company debate surrounding it.

It's uncertain when workers first began pursuing this strategy, though by the 1940s a mine inspector, Noronha Feyo, was actively railing against it, thereby raising its profile in the written record. His first report related to this practice was filed in August 1946, after a visit to Cassiaxima mine, run by a Sr. Barracosa. In it, Feyo noted that "workers do not eat their noon-time fish. . . . Sr. Barracosa replied that 'Some of them eat it in the afternoon in the village, others sell it or trade it or give it to their families.'"[15] In response, Feyo gave Barracosa instructions to oblige the workers to eat it in his presence and to prevent them from leaving the cafeteria until they had consumed their entire meal.

Less than a week later, Feyo noted that on Mugigi 2 mine, Sr. Silvino Balbino, the mine boss, was actually enabling this practice. According to Feyo: "Workers receive all of their fish and cornmeal for their evening meal at lunch. And, Balbino allows the [contratado] women to come to the exterior of the cafeteria in order to receive the meat or fish ration from the workers."[16] As the practice of distributing daily rations to the wives of contratados undoubtedly lent itself to this redistribution of workers' mid-shift meals, just a month later Feyo was again addressing similar events on Maludi 6 mine:

> The wives of the contratados descended as a group upon the mine, and under the pretext of bringing "tidbits" to their husbands, carried away part of the midday ration: I saw *pretos* [a pejorative term for blacks] transporting to the village part of the fish, beans, and even the manioc; I found other women a few meters from the mine eating the workers' rations right there. Because it was essential to put an immediate end to this practice, I called the mine bosses' attention to this occurrence. The contratados' wives, starting tomorrow, will no longer be authorized to come to the mine.[17]

Feyo proceeded to travel around Diamang's mines leaving in his wake a trail of these corrective instructions. Mine workers, and especially their wives, who had been the direct beneficiaries of this "redirected" food, naturally greeted them with hostility. A sequence of events stemming from a September visit to Luxilo 1 mine run by a Sr. Magalhães Queiroz brought this response into sharp relief for Feyo.

In his report, he outlined the initial incident as follows: "It was the day of the meat ration. At the end of the meal, I noted that the majority of the workers left it untouched or only ate a small part of it. Queiroz explained to me that on the day that meat is served it is always a struggle to make the workers eat it; they invent all kinds of strategies in order to take it home with them to the village to give it to their wives. Ordered to swallow it, a certain level of protest could be heard."[18] In this instance, though, Feyo's directive also precipitated a reaction from the workers' wives. "I have been informed . . . that the wives of the contratados have begun protesting, here and there, against the fact that the company obliges the workers to eat the ration in the mine cafeterias, and especially those wives who are now lacking the meat that the workers used to save for them. . . . [As a result] rumors are currently circulating of reprisals by the women related to the cultivation of crops if the current system is maintained."[19]

It is unclear what transpired next or what actions these women, their husbands, Feyo, or even Queiroz took, if any. However, it is revealing that shortly afterward, Diamang moved to circumscribe workers' employment of this strategy by uniformly issuing rations for families on a weekly, rather than daily, basis, thereby reducing contact between contratado husbands and their wives during the day. Increased rations for family members soon followed, thereby removing the original and ongoing primary impetus for workers to redistribute their vital midday meals.

A CULTURE OF COMPLAINTS: AIRING GRIEVANCES,
SEEKING IMPROVEMENTS

Diamang's African workforce regularly issued complaints to company officials and mine inspectors about insufficient rations, abusive treatment, and excessive labor demands. In this sense, the "culture of complaints" that workers cultivated places the levels of efficacy and tolerance of laborers' grievances at Diamang somewhere between African mining contexts in which labor unions formally articulated workers' complaints and those in which mine workers either had little faith that management would respond or, worse, would respond violently.[20] Even on the Angolan mines, this strategy constituted a calculated gamble, as complainants risked corporal punishment for their perceived audacity. More often, though, Diamang officials receiving workers' objections either simply ignored them or acted on them. When African employees' complaints were effective, they prompted a series of company actions that benefited both the workers who had issued them and their more reticent colleagues.

In the early years of the company's operations, workers' complaints about rations were the most prevalent. In 1925, for example, laborers took advantage of a visit to Diamang by Angola's governor, Bento Roma, to complain that food provisions were insufficient. In response, company officials accused these employees of prevaricating and countered that African employees were receiving even more than the law required. Ultimately, these complaints pushed Roma to order an alternative allocation of rations that increased the daily manioc flour ration from 300 to 350 grams, but reduced the weekly ration, issued on Saturdays, from 5,200 to 4,900 grams.[21] Later that year, workers passed along a similar complaint to E. Torre do Valle, the chefe of Fronteira do Chitato posto, during his visit to Luaco mine. In this instance, laborers complained that when they received meat "it is just the bones, while the whites get all the meat."[22] In fact, before the improvements of the 1930s, company records are replete with complaints related to the inadequacy or poor quality of rations.

Grievances related to rations declined dramatically in the 1930s and then more gradually over the ensuing decades, reflective of the company's progress on this front. By the 1950s, Diamang appears to have satisfactorily addressed rations-related issues, as workers only occasionally complained about rations, typically regarding their (inadequate) size. Beyond this decade, complaints related to food allotments disappear entirely from the written record, while informants indicated that rations were generally sufficient, if often (still) lacking in variety. Regardless, laborers' complaints had been an impetus for Diamang's improvements in this area, as the company wanted not only to retain existing workers but also to attract new recruits and eliminate inadequate food distribution as a potential source of aversion.

Over the course of Diamang's operations, the number of African employees' complaints concerning physical abuse outnumbered rations-related grievances and failed to ebb as dramatically over time. When workers felt that mine overseers' behavior was egregious, they often complained, hoping to effect a return to "normal" conditions.[23] A 1927 incident highlights workers' employment of this strategy, but it also underscores the potential risks. In this case, a contratado named

Masseca had reported a case of abuse, which the company recorded in the following manner:

> Masseca was working as a contratado on Cavuco mine, whose boss, Weatherby, known by the name of "Lunganga," began treating the workers very badly following the last visit by the governor of the district, attacking . . . those who had complained of the deficiencies in the nourishment they were provided. In virtue of the bad treatment they consequently received, five workers had already deserted, but Masseca, having been slapped and kicked by "Lunganga," did not want to flee to his homeland, preferring instead to present his complaint to the authorities. I sent the complainant back to work, and he promised to comply, guaranteeing to him that measures would be adopted so that the chefe of the mine would be more sensible with the workers.[24]

It is unclear whether or not Weatherby further punished Masseca after the latter had lodged his complaint, or if the mine boss honored the pledge made by the official who recorded it; regardless, the incident illustrates how workers had access to an array of strategies, including issuing complaints or deserting, to achieve their particular objectives.

Laborers also registered several complaints directed at ethnic Luba capitas (mine overseers). In fact, workers' initial complaints about these overseers had prompted Governor Bento Roma's 1925 decree that called for the replacement of all non-Angolan, that is, Luba, overseers (see chapter 2).[25] In another incident from the 1920s, a capita (of unknown provenance and ethnicity) notoriously made workers hunt rats to augment his standard rations. After workers complained, the company ordered the capita to serve thirty days of correctional labor.[26] Costa Chicungo, who began at Diamang in the 1950s, confirmed that reports of this nature persisted. "It was possible to complain . . . about conditions on the mines, and specifically the physical abuse by capitas. . . . The company would order the offender to appear, and if it concluded that he was a "big problem," then he would be fined. But if the offender had killed someone, then he would be imprisoned by the state."[27]

In other instances, capitas were themselves victims of physical abuse and consequently leveled their own complaints. In a case from 1930, a capita named Cajama complained about the abuse he had received at the hands of the white supervisor of Cassanguidi mine. The proceedings were summarized as follows:

> The white agent of Cassanguidi mine used a whip made of an automobile inner tube and hit Cajama in the hind, on the hands and at various places on the feet and chest. Cajama says that the white used to beat people regularly with it. The capita was beaten because he knocked some mud down on the mine that smothered another person, though not fatally. . . . Other Africans testified during a subsequent investigation by the chefe of Fronteira de Chitato and all declared that the capita was beaten severely and that the white had beaten almost all of the workers at the mine with the whip, that a day rarely passed where he did not beat one of them, that the said complainant had been beaten before, and that the white rarely had reason for these beatings, or he would justify it by saying that the Africans were working in bad faith.[28]

In most cases involving abuse, capitas were neither victims nor victimizers, but simply bystanders unable to help African co-workers address their grievances. A former capita, Alberto Rossa, explained: "Very few complaints were made to me. Some that were voiced concerned physical abuse, but I could not do anything about it because the ones who had the mission to protect the indigenous were the chefes do posto and they were complicit with the company and those committing the abuse. So, the colonial authorities were not complying with their own charge and even forced the Angolan to obey."[29]

By the 1960s, company records indicate that workers were articulating fewer grievances than in years past, while informants offered mixed testimony concerning the prudence of issuing complaints during the closing decades of Diamang's operations. For example, when I asked Anna Maria dos Santos if workers in the 1970s ever complained about conditions at Diamang, she succinctly replied, "A onde [to where]?"[30] Conversely, Felipe Leo Muatxissupa, who began working with the company in 1966, indicated that his articulated dislike for shoveling overburden ultimately landed him a job at a lavaria. "I worked for three years with a shovel and later at a lavaria. This work was better; shoveling was very tiring. In order to get this job, I simply explained that I did not want to shovel anymore and then proposed that I be transferred and my boss said, 'yes.'. . . I continued on at the lavaria all the way until independence. . . . Sometimes people would be scared and would not complain or ask for different work, others were not."[31] Although Diamang may have closed off this strategic avenue for some, continued evidence of laborers lodging oral, or even epistolary, complaints and requests highlights both their ongoing commitment to actively improving their work environments and their belief that (most) company officials were at least willing to listen.[32]

UNCERTAIN RETURNS: WORKERS' HIGH-RISK STRATEGIES

In addition to engaging in activities that entailed relatively little, or even no, risk, some male employees also pursued strategies that subverted company policies and therefore carried more severe consequences. These measures included withholding their labor, seeking medical absences on dubious grounds, or even engaging in diamond theft. Because these strategic pursuits compromised either production or profits, or both, over time Diamang moved to curtail them by expanding its supervisory and vigilance staffs while also addressing the original impetuses for these undertakings. With the passing decades, the combination of these company measures and African laborers' escalating professionalism resulted in a sharp decline in workers' more drastic endeavors, though employees never completely abandoned them.

GIVING IT LESS THAN THEIR ALL: WORKERS
WITHHOLDING THEIR LABOR

Over the course of Diamang's operations, African employees periodically engaged in concerted work slowdowns or, more often, individually took unscheduled or extended breaks, or simply gave less than their all. If capitas or European overseers detected or even suspected any of these measures, they subjected the accused to corporal punishment, though these strategic actions typically transpired undetected. Over time, the company countered a worker's potential desire to withhold his labor by establishing the aforementioned tarefa system at selective sites and, even more potently, by improving overall conditions on the mines.[33]

Workers began withholding their labor either individually or concertedly from the company's operational inception.[34] Because Diamang never instituted any type of bonus system, laborers had little incentive to transcend either what they perceived the minimum standards to be or the official targets following the implementation of the tarefa system. Consequently, company records from the 1920s abound with instances of officials accusing workers of exhibiting substandard effort. For example, in December 1927, Diamang officials were livid at two African workers, Chitambala and Sachimboio, who had been charged with cleaning out a new stretch of road between the Chingufo and Chimana mines, but who had "in three days, cleared only six meters, as they spent the majority of the time sleeping in the grass."[35] In September 1928, E. S. Lane, a Diamang official, compiled several similarly "flagrant" episodes that allegedly underscored workers'—and specifically contratados'—"bad will":

> In mine no. 3 of the N'Zargi mine group, a rate greater than 50m³ [cubic meters of cascalho removed] has not been achieved, although we'd hoped to reach a rate of 70m³. It has been noted that the workers that handle the wheelbarrows abandon them immediately after the European that oversees them turns his back. . . . These workers fill the wheelbarrows only half, or a third. . . . Those who work in the removal of the overburden layer are satisfied with the removal of 2m³ per day, when other workers, working beside them, remove 6m³ in the same amount of time. There are even indígenas who, asking permission to excuse themselves with the objective of satisfying certain bodily needs, spend two or even three hours . . . before returning. . . . These types of acts have affected the production for June and July. . . . The current situation is highly prejudicial to the common interests of the company and the state.[36]

A follow-up letter from Parkinson to Lane suggests that laborers were employing these tactics strategically, purposefully exploiting less experienced overseers. "Since the governor's visit, things there have gone from bad to worse. Since Agent Calçada left they have been trying out their tricks on Agent Remacle, who does not know what to do. The situation now is that they know they can loaf on the job in defiance of us. We have no control over them whatsoever and are unable to get a day's work out of them."[37]

Over the ensuing two decades, company revenues, which underwrote both increased work-site vigilance and corporate paternalism, coupled with laborers' ascending levels of professionalism, resulted in workers pursuing this strategy much less frequently. Moreover, the implementation of the tarefa system on select mines effectively invalidated it, especially as most tasks, from the removal of overburden to the number of diamonds "picked" each day in selection stations, were quantifiable.

This measure did not, however, cut off workers' access to these types of strategies entirely. Informants indicated that, even if only rarely, laborers continued to engage in work slowdowns during the 1960s and 1970s. Costa Chicungo, who was with Diamang from 1952 to 1974, declared that employees, typically in small groups of four or five, occasionally organized slowdowns, especially on mines without task-based systems.[38] Similarly, Luciane Kahanga, who was at Diamang intermittently during the 1960s and 1970s, indicated that when a capita identified a laborer not working at the "proper" pace, the overseer might escort the worker to a colonial administrative post where he would be beaten. Kahanga's testimony confirms that, despite the potential repercussions, a small number of workers continued to employ these types of strategies, even if over time the company's countermeasures greatly reduced both their allure and efficacy.[39]

MEDICAL ABSENCES: A STRATEGIC EXPLOITATION OF HEALTH

Contratados angling for temporary, or potentially even permanent, relief from the work regimen often sought a respite on medical grounds.[40] This strategy was an effective, though risky, undertaking that could provide one or multiple days of reprieve. Workers garnered approved medical absences by feigning illnesses, self-inflicting wounds, or intentionally neglecting minor injuries. Although mine managers generally displayed skepticism toward laborers who, they perceived, were seeking medical absences on dubious grounds, Diamang's Health Services personnel encouraged mine bosses to accept that at least some of these claims were sincere. As increasingly informed company officials better understood the range of diseases and injuries that afflicted the African workforce, ever more space was available for laborers to employ this avoidance strategy. Contratados still pursued it at great risk, though, because just as Health Services staff were better able to diagnose illnesses and their respective symptoms over time, they were also better able to detect insincere claims. Furthermore, mine overseers often remained the ones who determined—accurately or otherwise—the legitimacy of workers' assertions. And if either these mine bosses or Health Services personnel concluded that a worker was acting in bad faith, he would most likely be subject to physical punishment. Moreover, the self-infliction of a wound or the intentional neglect of an initially minor injury in the hopes of being temporarily excused from work, or even "repatriated," were actions fraught with obvious risks.

In the initial decades of Diamang's operations, company officials credited contratados with seeking medical absences in a variety of creative ways, with the latter's ultimate "prize" being repatriation on medical grounds. For example, in 1930, the head of the company's Health Services division, A. A. de Almedia e Souza, accused contratados of "hiding injuries [e.g., cuts and ulcers] until they get so bad that when they finally report them they know they will not have to work or that they may even be repatriated."[41] As evidence, he argued that as the number of contratados rose, so too did instances of untreated injuries, and thus he opined that "the contratado—unlike the voluntário—is the sort who detests work and disregards his physical integrity, even trying hard to compromise it with the objective of achieving repatriation."[42] It appears that sentiments emanating from the field, such as those held by Almedia e Souza, were cogent enough to convince the director, Ernesto de Vilhena, that contratados were pursuing this strategy to avoid work either temporarily or permanently. Consequently, Diamang's director asserted in 1929: "The voluntário not only works more and with greater will, but cares for himself, does not frequent the infirmary, avoids everything that can damage him physically—the opposite of the contratado who injures himself on the job, even deliberately—or infects the wounds and ulcers to which the natives are very susceptible, with the intent of being hospitalized and, thus, filling up, with the least effort possible, his contractual period."[43] Company medical records confirm that contratados visited the hospital three to four times as frequently as voluntários, suggesting that there was some truth to these accusations. Of course, the fact that contratados typically worked under more demanding and hazardous conditions undoubtedly contributed to this pattern.[44]

Diamang's insistence on ruling whether accidents and any consequent injuries incurred were intentional offers further evidence of workers' pursuit of this strategy.[45] In fact, the diamond enterprise was so positive that contratados were willing to deliberately sustain injuries to avoid work that they investigated for any evidence of premeditation, even in cases such as Chumanga's (outlined in chapter 4), who had to have his right hand amputated following an occupational accident.[46] Although determinations of intentionality were also required for compensation and pension purposes, senior company officials' pervasive skepticism suggests that they were concerned with more than just minimizing potential payouts.

Although Diamang could do little to prevent workers who were determined to injure themselves from doing so, it did try to reduce the number of laborers who consciously allowed minor injuries to worsen in the hopes of securing a medical absence. Although by the 1970s instances of workers pursuing this strategy were extremely rare, a company document from 1971 spelled out both the problem and Diamang's solution:

> Some ailments that cause long stays in the hospital were originally only small injuries. These are attributable to the carelessness and short-sightedness of some workers, who minimize these injuries and at times even try to hide them. . . . At lunch time, the mine boss and mine nurse will inspect all the workers, registering in a book the presence of those with injuries. These workers will be placed in a separate formation and will receive a number. . . . Only after this will they be served the meal. All workers with injuries will go to the first-aid post to receive treatment. . . . Before leaving the mine, the boss will verify if all those requiring treatment received it.[47]

As with many company initiatives, discrepancies between policy and practice existed. Mine bosses were rarely this diligent, nor were nurses or first-aid posts always present. However, Bernardo Montaubuleno, who worked for Diamang's Health Services from the early 1950s until 1975, could at least confirm that "when company bosses saw that injuries had become infected, they would send the worker to the hospital."[48]

Other employees sought medical absences by feigning illnesses, which though painless and innocuous compared with self-inflicted wounds or neglected injuries, could elicit corporal punishment if attempted unsuccessfully. The first mention in Diamang's records of a worker pursuing this strategy was related to the 1927 incident described earlier in this chapter involving an attack by a mine overseer named Weatherby, or "Lunganga," on an employee named Masseca after the latter had issued a complaint. In a subsequent letter, Weatherby rebutted Masseca's account of the incident and accused him of feigning illness, claiming that only after a company doctor had determined the spuriousness of his alleged ailment did Masseca go to the

Chitato administrative post to complain, rather than to Cassanguidi mine where the doctor had directed him to proceed.[49] This insubordination, in turn, served as Weatherby's justification for the abuse, which he deemed commensurate with an offense of this nature.

In later years, informants indicated that feigning illness remained a strategic option and continued to carry the same risks. Alberto Rossa, a former capita who began at Diamang in the 1940s, indicated that "contratados would often feign illness. If they were discovered to be lying, they would then be beaten and sent back to complete their original task."[50] Bernardo Montaubuleno, who joined the company in the 1950s, echoed this affirmation, recalling: "If the company was suspicious of an attempt to feign illness, the worker would be taken to the cafeteria in order to confirm or deny it, and if he actually was [ill], he would then be sent to the hospital so the doctor could perform a more thorough examination. If he was found not to be [ill], then he would be sent to Cambulo, the administrative post near here, to be punished. He would be struck on the palms with a palmatória."[51]

RISKING IT ALL FOR A STONE: DIAMOND THEFT

Over the course of Diamang's history, a very small number of African employees opted to steal diamonds from the work site. Laborers who worked in areas such as washing or selection stations, at which contact with diamondiferous ore in advanced stages of refinement was a daily occurrence, were the most likely to engage in theft. This activity was extremely risky, as workers faced physical abuse, jail time, fines, exile, or even capital punishment if caught. Those who succeeded sold their contraband to white employees, sobas or capitas, who then tapped into far-reaching networks to smuggle the stones out of Lunda to destinations around the globe. In the orbit of this activity, the company and individual practitioners continually tested one another, with suspicious officials regularly instituting reactive security measures to address systemic openings that workers had been exploiting.[52] Unfortunately, it is impossible to quantify or even accurately gauge trends related to diamond theft because pilfering diamonds was by nature a clandestine, or hidden, pursuit. However, oral evidence and arrest rates suggest that over time company countermeasures greatly reduced, but were never able to stem completely, the outflow of these illicit stones.

In Diamang's early years, officials reasoned that the company needn't concern itself with African employees because these workers allegedly didn't understand the "true" (market) value of diamonds.[53] Parkinson, for example, declared that "at the beginning of operations in Angola, most of the natives did not have any idea of the worth of a diamond, which was left lying around until it was turned in at the end of the shift."[54] Further, according to H. T. Dickinson, a Diamang engineer, "When I was at Maludi mine in 1922, I was very unfavorably impressed by the absolute lack of precautionary measures against possible diamond thefts, this being, however, a natural consequence of the conviction entertained by Mr. Newport [the managing director in Lunda] that the black workmen did not have the least idea of the value of the stones that they were employed to extract."[55] Despite Dickinson's unflattering review of company security, Diamang officials' naïveté did not last long. By the mid-1920s, they were well aware that some employees had been taking advantage of this laxity and that the enterprise had to take measures to address the most blatant security gaps.

According to company officials, early diamond thefts were carried out by African employees who, even if they knew little of a stone's market value, could readily sell them to someone—African or European—who was willing to pay them handsomely for their efforts.[56] Mine bosses mainly suspected "pickers," but also those laborers who were exposed to unrefined cascalho, including those still using baskets to transport the gravel to washing stations. Diamang claimed that individuals involved in theft would "sell diamonds for themselves, to their sobas or directly to whites . . . who would, in turn, transfer the stones to other individuals, nationals or foreigners, interested in the illicit traffic. By these hands . . . the stones traveled to the metropole or to other countries, so that, from investigations already undertaken, it has been concluded that entities from Lisbon, Antwerp, and Rotterdam are involved in such thefts."[57] Consequently, the company accused sobas, white settlers and even capitas of pressuring mine workers to steal stones, seemingly failing to consider that the African employees who engaged in this activity also benefited and, therefore, may very well have initiated it themselves.[58]

Once Diamang officials acknowledged that diamond theft was a problem, suspicion increasingly colored their view of the African workforce. A company document from 1928 reveals that company officials suspected some Africans of "showing up asking for work either intending to steal and sell diamonds for themselves or as emissaries of whites involved in the illicit trade."[59] In fact, even in the absence of any evidence, Diamang officials could convince themselves that a crime had occurred. In 1936, for example, a senior company official, Osório Júnior, was certain that a worker named Sachai, who had deserted, was involved in illegal diamond activity. Júnior wrote: "Sachai's flight was suspicious because it was not due to poor treatment, or an incident, or because of an illness, and also because he was almost done with his contract and had missed only one day of work thus far, and now he was abandoning his salary reserves, which came to a sizable sum for an African. I suspect he was one of those Africans who, incited by someone, goes to work on the mines with the thought of robbing diamonds and abandons his work immediately after he has obtained them."[60]

Validating company officials' myriad suspicions were the arrests of African employees (as well as whites, away from the mines), which Diamang was regularly making by as early as the end of the 1920s, for diamond theft or participation in the illegal trade of stones. At the end of 1928, for example, the company claimed to have recovered 8,195 carats worth of contraband diamonds from various arrests and seizures.[61] African employees who were caught faced stiff penalties, including fines and imprisonment, as well as severe beatings. For example, a 1931 company report revealed that "various prisoners continue to die in the prison of Cambulo, some of them thieves of stones. In addition to Sota, who stole two diamonds on Lussaca [mine], Iamba-Amba, who robbed five stones, supposedly diamonds, from the mine of Cavuco, also died."[62]

By 1932, six Europeans and fifty-four Africans were in prison for diamond theft and commerce-related offenses, with most of the latter serving eighteen-month terms.[63] Meanwhile, the colonial state was also stiffening penalties for any perpetrators it apprehended, including deporting offenders to the islands of São Tomé to perform "corrective labor." In 1946, for example, the state exiled eight men to this island colony, and in the preceding four years the figures were three, two, twelve, and six, respectively.[64]

Into the 1940s, African employees continued to remove diamonds from Diamang's installments, albeit in small numbers, as evinced by both arrests made and stones recovered (9,402 carats in total by 1947). In 1946, the arrest for diamond theft of a "double agent" named Sucasuca, who had been working for Diamang as an informer and had helped the enterprise recover diamonds stolen from Maludi mine in 1944, generated an embarrassing moment for the enterprise. A January 3, 1946, company memo proclaimed: "It has been discovered that Sucasuca was leading a gang of diamond traffickers. The gang is all imprisoned, but he is still being sought. . . . He was apparently planning on filing a false report in order to obtain a reward from the company, presumably after producing or leading the officials to the stolen diamonds, which had been buried in a marsh."[65] Shortly thereafter, on January 9, Suca-suca was apprehended. Ultimately, his case underscores the degree to which some African laborers at Diamang were willing to go to enhance their lives.

Throughout the 1950s, at least some workers continued to try to capitalize on their employment at Diamang by stealing diamonds. During this decade, small numbers of Africans were arrested—often those employed at lavarias who had removed stones from the highly refined concentrate. In 1957, however, the company uncovered an organized ring of African employees after some members were caught stealing diamonds from the Central Selection Station at Andrada. These individuals subsequently informed authorities that they had African accomplices in the neighboring Belgian Congo.[66]

Arrests of African employees continued into the following decade. In 1969, for example, a high-profile case involving a laborer working at a lavaria attracted a great deal of interest. After he was caught and interrogated, the company discovered that

> a few days ago, the fifteen employees from the lavaria met . . . and arranged to steal diamonds. . . . They agreed that the most propitious moment would be during the washing, . . . an operation that precedes the selection. As they knew an interested party involved in the sale and commerce [of stones] from this area, one of them, Alfredo Xaissambo, entered into a contract with him, fixing the date and the location for the transaction. . . . Days later, he was approached by an individual from Malange, a mulatto, strong and short, who was to buy the diamonds, allegedly for 500 escudos total [fifteen diamonds, one from each conspirer].[67]

The apprehended employee further revealed that they had been able to steal the diamonds simply by distracting the capita watching over them. Consequently, in spite of Diamang's success in greatly limiting theft of this nature, a South African geologist who was visiting the company's installations during this period declared that "at Diamang, there was so much richness and stones so grand, contrasted by so little security."[68]

By the 1960s and 1970s, the company had markedly heightened worker vigilance, especially at the selection station. Yet despite all the new measures, including technological innovations, introduced to prevent theft, a handful of employees continued to pilfer diamonds. Leonardo Chagas, Diamang's former security chief, chuckled as he recalled, "Workers at the selection station were the biggest culprits. They would stick stones under their tongues after creating a small distraction. . . . Later on, we made them undress and checked to see if they had anything in their body or in their anus. Despite this, they were always trying."[69] This pervasive suspicion pushed other workers, including even those who were in no way involved in diamond theft, to employ precautionary measures. For example, Paulo Leão Vega proclaimed: "We would not use the word 'diamond' when talking about them, but rather '*ginguba*' [peanut] as a code word so that the Portuguese could not understand."[70]

Informants cited the acute levels of vigilance and the severe consequences as reasons for eschewing larceny while at Diamang. In fact, most of the former employees whom I interviewed vehemently insisted that diamond theft quite simply did not occur, owing to potential prison sentences, or worse. For example, when I asked about the possibility of getting involved in the illegal diamond trade, Vega replied in the same fashion that many other informants did: "*Nunca, nunca, nunca* [never]. You would be killed."[71] Andre Muamukepe, who was with Diamang from 1954 until Angolan independence, echoed these sentiments: "You would never be seen again if caught with diamonds."[72] Similarly, Isabel Reis, a Portuguese who grew up in Dundo, grimly explained that if an African employee was caught stealing diamonds, "He might have been beaten; he might have been put at the border; he might have been thrown in the river and he might have been thoroughly examined. There are all kinds of exams you can perform on a body."[73] While Diamang's paternalistic measures were sufficient to keep most of workers from engaging in diamond theft, the company did not hesitate to resort to extreme violence when, in rare cases, these proved to be inadequate.

NO-SHOWS: ABSENTEEISM AND DESERTION

Over time, and in particular during Diamang's early operational years, many African employees were not content to attenuate the daily labor regime or even to stay within the system to resist or subvert it, but instead elected to completely remove themselves from it. To withhold their labor in this fashion, workers simply failed to report for work. When voluntários stayed away from the mines, the company labeled it absenteeism. Officials disliked the unpredictable disruptions this strategic practice caused, but they tolerated it to an extent in the hopes of amassing an exclusively volunteer workforce composed of experienced, proximate, and (relatively) inexpensive laborers. Conversely, when contratados failed to arrive at work, the company immediately identified these individuals as deserters, and subjected any apprehended offenders to corporal punishment. Over time, as Diamang consolidated regional power and upgraded conditions on its mines, absenteeism and desertion became both decreasingly viable and appealing, while productivity and production levels correspondingly increased. Indeed, it is telling that none of my informants had engaged in either absenteeism or desertion, or personally knew anyone who had.

VOLUNTÁRIOS AND ABSENTEEISM

Since voluntários engaged with Diamang via loose, often oral, contracts, or without any formal commitment at all, when they failed to show up for work mine bosses defined this behavior as absenteeism rather than desertion. Many voluntários withheld their labor in this manner not only to avoid the demanding work regimen, but at other times simply to attend to matters at home, such as tending to crops or an ailing family member, to hunt or even to avoid commuting during heavy rains. These unanticipated absences disrupted Diamang's scheduling and task assignments. Yet because the company so desperately wanted to attract a permanent, voluntary workforce, officials often reacted with disapproval rather than violence. Therefore, this strategy constituted a low-risk endeavor for voluntários, as temporary or permanent dismissal was a much more likely outcome than corporal punishment. But as Diamang only paid these casual laborers for work performed, it also had a considerable downside.

Isolated episodes of absenteeism began appearing in company records as early as the 1920s. In the last two months of 1923, for example, Diamang lost several voluntários who had returned to, or remained in, their villages at the beginning of the planting season. When company officials attempted to send voluntário laborers from Dundo to replace them, "The attempt was a failure because the workmen returned to their villages and would not go to the mines. . . . In fact, the only result was a reduction of the number of laborers in Dundo."[74] A 1929 incident in the Chingufo mine group highlights a similar display of worker intransigence. In this case, as operations in this area ended and the transfer of equipment to Andrade mine began, E. S. Lane, a Diamang official, wrote: "We thought the voluntários in service in Chingufo should go on to Andrade as, from the beginning, they had promised to do. However, at the point when this transfer was to occur, these indígenas declined to go to Andrade and, as such, we had to substitute contratados for them."[75] Although company officials offered no explanation for this uncooperativeness, it's likely that in both instances voluntários elected to withhold their labor because they were reluctant to relocate away from their homes and families.

As mining workforces and operations began to stabilize in the 1930s, company officials increasingly bemoaned any occurrences of absenteeism. Dickinson noted in February 1931, for example, that "the use of volunteer labor in place of contract men is certainly beneficial, but at the same time we are having more trouble with irregular attendance of volunteers at the Cassanguidi and Andrada mine groups than we have experienced with contracted men."[76] Expressing similar concern less than a year later, Parkinson informed de Vilhena that due to voluntários' rampant absenteeism, "We were eventually forced to ask for 500 additional contracted men from Saurimo."[77] In both of these cases, company statistics support officials' complaints. From 1931 to 1933, contratados had an average attendance rate of 80 percent, while voluntários' rate was only 67 percent, and had been as low as 54.96 percent in 1930.[78]

Beyond depriving Diamang of valuable working hours, absenteeism also severely disrupted the company's ability to schedule tasks for laborers, causing the enterprise to incur additional expenses. A 1931 letter from H. J. Quirino Fonseca declared: "The voluntário is irregular and less reliably present compared to the contratado. . . . In order to address these absences we are obligated to have a larger contingent of workers than is necessary, which represents a financial burden to us. The contratado laborer, however more costly to the company, appears more regularly for his job and is by nature more stable."[79] In its 1935 annual report, Diamang applied hard numbers to this generalization and determined that to guarantee that 100 voluntários would be at work on a given day, it was generally necessary to have 210 under some form of contract.[80] Meanwhile, that same year, contratado attendance rates had risen to 94.3 percent.

The company tolerated, if only begrudgingly, the high levels of voluntário absenteeism for several reasons. First, contratados were, as Fonseca mentioned, more expensive. In 1940, Diamang estimated that the average per-day cost of a contratado was 106.7 angolars, against 73.2 for a voluntário, and these figures had risen to 364 and 191.3, respectively, by 1960.[81] Second, company officials consistently envisaged an eventual phasing out of contratado workers, which meant it had to manage carefully its relationship with the more proximate voluntários. And, third, as a 1937 memo outlined: "Voluntários, who have lower attendance rates than contratados and work for only three or four months at a time, often return for additional stints, at which point their productivity rates increase, or are already high, because of their familiarity with the job."[82] Over time, Diamang's patience, tolerance, consolidation of local authority, and pragmatic, paternalistic upgrades paid off: in 1940, it was necessary to have only 143 voluntários under some form of contract to ensure that 100 would be at work on a given day, while voluntário attendance rates rose steadily and by the 1940s were consistently over 90 percent.[83]

Elevated voluntário attendance rates, however, did not entirely mollify mine managers. These supervisors continued to struggle with both the uncertainty that absenteeism bred and the ongoing absences themselves, which, although decreasingly common, could still, on occasion, be disruptively high. For example, an August 13, 1949, inspection of Mufo mine revealed that 33 out of 257 voluntários were absent. The mine inspector, Santos Ribeiro, noted: "The special situation of this mine, in which the tasks are mainly performed by voluntário laborers from the region of Canzar, who are here . . . for just a short period of time, makes it difficult to assemble good teams for each task, and the low attendance rate of these same workers aggravates this difficulty."[84] As the labor supply stabilized over time, however, individual mine supervisors typically dismissed chronically absent workers rather than tolerate copious absences.

By the 1960s, laborers' absenteeism had waned considerably. For decades, attendance percentage rates for voluntários had been hovering in the mid-90s. During this decade, however, the company lost ground in its ongoing battle to combat any lingering absenteeism due to the introduction of colonial legislation in 1963 that partially legalized the practice. This update to the colonial work code allowed workers three excused—and thus paid—absences per month for specific reasons, such as a death in the family (a recurring reason for at least some unannounced worker absences). Despite protests, albeit rather tepid, by Diamang officials, the company eventually agreed to uphold the new legislation. Anna Maria dos Santos, who worked at Diamang from 1973 to 1975, explained that "after three absences in a month, you would be fired, but you could ask for time off for a relative's death, or something like that, and it would be no problem."[85] Diamang also appear to have softened its stance toward absenteeism, perhaps as a result of this legislation. In 1972, for example, the manager of Saga mine had granted "twelve workers time off . . . in order to relocate their villages and families, currently in the area of Lóvua, to Saga. . . . Lately, this policy has been broadly adopted, as excused workers normally return to work in days, as opposed to not providing the exemptions, which had normally translated into these individuals not returning at all."[86]

Into the 1970s, voluntários, or "workers from the region," continued to accumulate unexcused absences, even if in exceedingly small numbers. Laborers' responses to a 1970 company survey shed some light on their motivations. Of the 1,409 workers who responded, 37.5 percent declared that they needed to help their wives with farming tasks or to repair their homes; 27.7 percent said they wanted to rest; 9.7 percent cited visits to relatives, sickness, "laziness," or a family illness; and other employees indicated that, inter alia, absences were due to excesses the night before or an unwillingness to walk to the mines when it was raining.[87] A letter from Diamang's director in Lunda to the administration in Lisbon illuminates the decidedly practical and tolerant approach the company had adopted "on the ground" to these very real motivations for worker absenteeism: "It's important to continue to honor the long-standing practice of workers alerting the mine boss a day in advance of when they will be absent because . . . it is important not to abruptly deny these workers the ability to attend to their [home] business nor should the practice be broken abruptly, in general. It is also advantageous to the mine boss, who then can plan around this absence and re-allocate labor to cover, if necessary, for the absence."[88] Meanwhile, absenteeism rates also generally rose when laborers were forced to resort to manual techniques following machinery failures. Given these inevitable, and steady, mechanical breakdowns, rates consistently hovered around 5 percent on into the 1970s, with occasional spikes but without any radical fluctuations.[89]

Contratados did not enjoy access to absenteeism as an occupational strategy. Company officials classified *any* unexcused absence by a contratado as an intention to permanently desert, immediately warranting arrest and/or corporal punishment. African employees who opted to proceed anyway were typically responding to insufficient rations or poor treatment on the mines, and thus high(er) desertion rates featured during the early decades of Diamang's operations. As with absentee *voluntários*, deserters also caused major disruptions for mine bosses who had been counting on their labor. Most successful fugidos quickly move out of the historical record, fleeing to the Congo, sympathetic villages in the region, or even daringly back to their home communities.[90] In the early years of Diamang's operations, when both the company and the colonial state enjoyed only limited regional hegemony and conditions on the mines were highly disagreeable, desertion was an appealing strategy for workers. From the mid-1930s onward, however, consistently low rates prevailed. It wasn't until the outbreak of the Angolan Revolution in 1961 that desertion was again thrust into the forefront of company officials' minds. Reports of fugidos assuming key roles in Congo-based Angolan nationalist movements granted the strategy a new, martial dimension.

During the first decade and a half of Diamang's operations, multitudes of contratados chose to flee, often causing the enterprise major fiscal and operational problems. A 1925 company letter expressing this frustration concisely describes how disruptive desertion could be: "Many contratados arrived on the mines in the morning and fled at night of the same day."[91] In 1931, desertions on the Cassanguidi and Andrada mine groups crippled operations to the extent that "it was necessary to close down some of the [individual] mines in order to get the best results from the limited supply of remaining workmen."[92] Company officials admitted that many workers deserted owing to inadequate, or even nonexistent, rations, and thus a certain level of empathy existed within Diamang's administrative circles.

Most of contratados who fled did so because of the poor treatment they received, including insufficient rations and the harsh work regimen that they were forced to endure. For example, after a group of fugidos was apprehended in 1929, these deserters claimed that "they did not want to return because they were beaten and made to work from sunrise to sunset, and that they wanted to go to the Congo, and that is why they deserted."[93] In a notable incident from 1932, three contratados fled all the way back to their home posto in Moxico—some 500 kilometers away—and subsequently complained that an epidemic on the mines had already killed four people, but that the company was not taking any preventative measures.[94] During this period, workers' concerns related to personal health and safety pushed desertion rates above 40 percent, at times, such as in 1930, while recapture rates remained in the low single digits.[95] Even when deserters were captured, there was little the company could do to prevent recidivism, even after a thorough hiding. For example, of the thirteen fugidos who had been recaptured in the first half of 1931, eleven deserted anew.[96]

By the mid-1930s, the impetus for, and strategic viability of, desertion decreased dramatically. As Diamang's profits steadily grew, its financial success enabled it to exert increasing pressure on local colonial officials to round up any fugidos and to stop blithely ignoring the reappearance of these deserters in their home villages. Meanwhile, the company was busy actively improving conditions on its mines. Collectively, these service upgrades, Diamang's solidification of regional power and the attendant increased state-company cooperation drastically diminished the feasibility and allure of desertion. In fact, this shift was so dramatic that company officials were dismayed when desertion rates climbed from 2.4 percent in 1938 to 4.1 percent the following year—rates that earlier in the decade would have been roundly celebrated (table 5.1). In this case, officials blamed the "spike" on the high(er) percentage of contracted workers who came from Moxico and Dilolo (south of Lunda), "regions that always supply bad contingents."[97]

From approximately the mid-1930s to the early 1960s, desertion rates remained stable, hovering between 2 and 5 percent irrespective of the numbers of contratados employed. These consistently low rates notwithstanding, officials strove to further thwart the practice as they

---

TABLE 5.1

Annual numbers and percentage of contratados deserting

	1930	1931	1932	1933	1934	1935	1936	1937	1938	1939	1940
Number	1038	1127	1093	279	63	294	373	301	225	390	271
Percentage	23.13	32.4	26.8	7.1	1.5	5.03	4.52	2.84	2.4	4.1	3.3

Sources: Manuel Pereira Figueira, *Relato sôbre mão d'obra indígena* (May 18, 1938), 39, MAUC, Folder 86 36°; Spamoi report (June 1941), 3, MAUC, Folder 86D 2°.

concluded that (successful) desertions begot more desertions. Even though officials were unable to completely eliminate this endeavor, when compared with the early years of mining operations it is clear that Diamang's countermeasures and initiatives were effectively limiting workers' pursuit of this strategy, as well as its appeal.

Although the numbers of deserters were extremely small from the mid-1930s on, those few employees who did abandon their jobs often reappeared elsewhere on the company's installations, strategically seeking less rigorous work. In 1930, for example, the enterprise was reporting: "Various deserters . . . have not been going back to their lands, but have been staying on at Diamang as voluntários, some in their same mines and others in Dundo as servants. When service on the mines, to where they've been sent, doesn't agree with them, they flee and will go offering themselves as voluntários in a capacity that is more agreeable to them, and they simply give a different name."[98] Similar reports were filed in the 1940s and even into the early 1970s, suggesting that many African workers were content with the prospect of remunerative engagement with Diamang, but (creatively) wanted to upgrade their occupational postings.[99]

In 1960, the neighboring Congo swiftly and somewhat unexpectedly gained independence from Belgium. Although this momentous occasion failed to spark an uprising or even any restiveness at all in Diamang's operational area, it did transform a proximate area into decolonized space, which appealed to certain Angolan laborers. Over the ensuing few years, company officials recognized that at least some of these deserters were involved in subversive activities across the border. And of particular concern was the exodus of Baluba, who had historic and ethnic links to the Congo, as some of these ex-employees were helping the newly formed Angolan revolutionary movements, whose primary goal was to topple the Portuguese colonial government.[100] On August 30, 1962, for example, company-paid informers in the Congo reported that a Luba fugido named Belebele Omer was commanding the troops of the fledgling nationalist movement, UPA.[101] Reacting to this new militaristic dimension of desertion, the feared PIDE initiated regional operations in Lunda. A 1963 report from the secret police organization revealed the depth and severity of this wave of post-1961-inspired desertion. "Many specialized [or skilled] workers fled to the former Belgian Congo from January to April of 1962 from Diamang. Also, some skilled Angolans from Luanda who sought work for

Diamang through the company's representatives in the capital . . . only stayed out in Lunda for a short time before fleeing to the Congo. . . . An informer in Dundo tells [us] how the UPA is encouraging Angolans to flee to the Congo and to support the UPA, especially former or current Angolans in the military."[102] Given the charged environment in the colony, Diamang was treating all episodes of desertion increasingly seriously, but none more so than those involving members of its diminutive, yet vital, skilled African labor force. Company officials understood well the value of these individuals to the movements assembling across the border.

Unfortunately, it is impossible to quantify workers' levels of flight subsequent to the developments of the early 1960s and on into the 1970s because Diamang stopped registering desertions following the dismantlement of the contratado system. Informants' hearsay and the spate of company- and state-generated letters related to desertion and security in the aftermath of the events of 1960 and 1961, however, confirm that at least some employees continued to pursue this strategy. Meanwhile, others continued to desert the company's operations for familiar—and much less sensational—reasons. For example, a 1972 letter from the field to Lisbon stated: "In order to get men from Calonda mine to relocate to the sparsely populated area where the new Lumboma mine is being put into operation, we promised them excellent housing, superior to what they currently have, based upon the expected arrival of [prefabricated] 'Trajinha' houses from the metropole. When the workers arrived, however, they found only old grass and wood houses, so they began to desert in large numbers."[103] This correspondence suggests that the African beneficiaries of the material upgrades that Diamang had begun introducing so many decades earlier continued to expect, value, and place significant importance on these ongoing enhancements.

PRODUCTIVITY AND ATTENDANCE: A POSITIVE CORRELATION

Declining rates of employee absenteeism and desertion generated corresponding increases in overall worker productivity, while Diamang's continual (re)utilization of experienced voluntários and contratados and its gradual mechanization of the mining process also escalated production levels. Unfortunately, it is impossible to disentangle this mélange of contributing factors when assessing annual carat output. Yet they do suggest that as African laborers increasingly exhibited occupational professionalism, the company responded in motivational, complementary, and reciprocal fashions, enjoying the resultant heightened productivity and profits.

The first spike in productivity occurred in the 1930s, coinciding with the broad-based improvements that the company had begun making. In December 1934, for example, the average number of cubic meters of cascalho removed per man, per day stood at 0.717; a year later, that figure had almost doubled, to 1.33.[104] By the end of the decade, this figure stood at 4.84 and had doubled again by 1953, to 10.22.[105] Innovations in the removal of these superficial layers, including the introduction of hydraulic processes, undoubtedly helped facilitate these gains, but workers' increasing levels of occupational commitment cannot be discounted. The company's growing investment in, and concern for, its African labor force is also salient. For example, in 1938, company officials commented that "there have been other . . . causes for the decreased output per man [over the past year], such as greater consideration for the health and welfare of our workmen," suggesting that Diamang was willing to endure short-term declines in productivity to satisfy its overarching, long(er)-term production objectives.[106] Periodic salary enhancements and bonuses also appear to have elevated productivity, and much more immediately. For example, following a series of mine inspections in 1972, undertaken following both a recent rise in salaries and the rare implementation of a bonus system for workers when attendance targets were reached, Fernando Duarte reported that "the workers told us they were satisfied with the increases in their salaries, [and] the mine bosses inform us that all the personnel under their watch have bettered, if possible, their output."[107]

⌐

Throughout Diamang's operational existence, African men and women engaged in a range of strategies to various ends. Employees most commonly employed low-risk, creative coping strategies that served to attenuate the harshest features of the daily work regimen but did not challenge the system in any significant or deliberate way. Many fewer engaged in more subversive acts that correspondingly featured both greater risks and, if successfully undertaken, greater rewards. Still others brazenly escaped the daily labor regime altogether, either temporarily or permanently.

Over time, the company's efforts to address flaws and interstices in its operational and security structures, coupled with improving conditions on the mines, prompted an increasingly "professional" African labor force to tailor their strategic pursuits. Activities such as singing and task sharing, which helped employees endure the workday while never threatening or undermining the labor process, remained consistently attractive to both mining officials and African workers. Conversely, endeavors such as diamond theft, absenteeism, and desertion plummeted as a result of Diamang's countermeasures and paternalistic initiatives, which elevated risk levels and removed much of the impetus, respectively, to engage in these activities. Yet even as members of the labor force increasingly eschewed particular strategies, they continuously and creatively adjusted their approach to the labor process, and thus at no point was the enterprise ever able to completely close off workers' access to these strategic pursuits.

Just as Diamang designed and instituted labor practices, procedures, and policies on its mines, and workers subsequently found space within them to engage in a range of creative activities, so too did this interplay occur after workers completed their shifts each day and returned to the company's mine encampments. In the following chapter, I examine life "after the whistle blew," which transpired in spaces removed from company work sites, and consider how laborers strategically interacted with Diamang representatives, nearby villagers who were otherwise unaffiliated with the diamond enterprise, and, most important, with each other.

6 ⤺ Eventful Evenings
Life after the Whistle Blew, 1925–75

With the aim of increasingly improving, both materially and morally, the living conditions of the African workers and their families, the Company organized in 1936, and has expanded progressively since then, the Secção de Propaganda e Assistência à Mão-de-Obra Indígena (SPAMOI). As a result of this action, not only . . . has the variety . . . of nourishment provided to the indígenas grown, but there has also been a very noticeable transformation in the hygienic conditions of their encampments, with the consequent improvement of the sanitary state of the workers and, in a general manner, of the entire population around the region of the mines. Just in this service, the Company presently employs . . . more than 600 indigenous auxiliaries.

—*Diamang Annual Report, 1937*

I am an Ambaquista, from Malange. . . . My father was contracted to work for . . . Diamang. He died . . . while we were there together. . . . By then, my mother was already dead. I cooked for my father on the mines and worked in the lavra during the mornings with other women. I learned to speak Chokwe while there. My father learned it on the mines, as well. There was no discrimination against those who didn't speak Chokwe. I . . . met and married a Chokwe man, João Avelino, on the mines. . . . We had to . . . get approval from the mine boss for a party for the wedding in the encampment. It was on a Sunday, but we still had to work on Monday! We were married in the Catholic Church in Maludi—at which I was a regular attendee. It had a racially mixed congregation. We moved in together after the wedding. Eventually . . . we had a child while at Diamang in the hospital.

—*Aida Fernando, who started at Diamang at fifteen years of age, in 1965*

DIAMANG'S CREATION OF SPAMOI (Secção de Propaganda e Assistência à Mão-de-Obra Indígena)[1] was intended to generate scenarios exactly like the one Aida Fernando conveyed to me during my interview with her (excerpted above). The company's "material and moral" initiative was directly linked to its provision of agreeable accommodations and popular post-shift recreational activities, as well as its achievement of food security, each of which helped to render the mine encampments sites of social cohesion. In these settings, residents such as Aida Fernando tended to ignore, rather than regard, an array of potential social fault lines, including ethnic, linguistic, marital, occupational, and tenurial. Testimony from Domingos Cazeweque, a former contratado who worked on the mines in the early 1950s, is emblematic of this "social professionalism." "I made friends with voluntários who lived near the encampment. We would buy and trade things with them. . . . There were also good relations between experienced and new workers in the camp. I learned a lot from them."[2] In fact, even as encampments grew more demographically diverse over time, residents "professionally" ensured that camaraderie, rather than hostility, characterized these spaces.

Although employees may have enjoyed the harmonious atmosphere in Diamang's encampments each evening, the range of challenges associated with the labor process did not conclude simply because the whistle blew. The daily commute home could reach a dozen kilometers and include river crossings or similar obstacles, thereby extending the labor process beyond the mines. On arriving home, retrieving potable water and preparing dinner constituted immediate priorities in this portion of the extended labor process. Diamang also partially monitored the activities in which men and women could engage in company mine encampments, including drinking (except on Saturday nights), and, at times, interaction with local communities, so as to inhibit the illicit trade in food, alcohol, and diamonds. Finally, wives of contratados could also find themselves subject to unwanted sexual advances in the company encampments by fellow residents, African sentinels (*sentinelas*), or European mine managers.

In response to these challenges, men and women engaged in several strategic activities to improve their "post-whistle" lives. Encampment residents brewed, sold, and drank alcohol, shared domestic tasks, cultivated and deepened an array of both platonic and amorous relationships, danced, sang, reproduced ceremonies and rituals, lodged complaints, socialized and traded with neighboring communities who were otherwise unaffiliated with Diamang, and, following 1961, listened clandestinely to the Angolan independence movements' evening radio broadcasts. These pursuits were generally no- or low-risk in nature (with the notable exceptions of brewing alcohol and tuning in to nationalist movements' broadcasts) and were essential in helping workers and family members enhance their lives while in Diamang's employ. Consequently, over time, company officials typically encouraged or simply displayed indifference toward these activities, reasoning that workers who were allowed to channel their energies in familiar pastimes and/or otherwise innocuous undertakings would be less likely to disturb the reigning tranquillity in the mine encampments.

Diamang also played an active role in generating and maintaining the stability that marked these mining encampments. From the 1930s onward, in conjunction with the creation of Spamoi, the company significantly improved housing conditions for workers and their families, offered arbitrative services to resolve intra-encampment disputes, achieved food security, introduced a series of extremely popular recreational activities, encouraged workers and any accompanying family members to attend (Catholic) church, offered educational opportunities (albeit limited) for workers and their children, and attempted to weed out abusive elements among its supervisory ranks. Although Diamang representatives often displayed violence toward encampment residents who threatened to disrupt the prevailing stability, company officials were generally responsive to workers' articulated grievances and, in general, endeavored to improve conditions "after the whistle blew" as part of the enterprise's broader paternalistic strategy.

In this chapter, I follow workers and family members as they returned home from work each day and examine the ways that both the company and the African labor force strategically cultivated the stability that characterized Diamang's mine encampments. For contratados, the commute home each evening brought them to these settlements, which typically housed between one hundred and two hundred workers and family members. Conversely, voluntários, who lived in their existing villages for much of Diamang's operational existence, experienced life away from the work site in familiar settings, well beyond the company's gaze. Consequently, the focus of the chapter is on contratados and their families, who continued to interact with company agents each evening, remaining under Diamang's aegis.[3] Throughout this examination, I seek to demonstrate that, over time, workers' and family members' strategic endeavors, combined with

Diamang's increasingly paternalistic policies and practices, dramatically improved the quality of life for Africans on the Angolan mines each day after the whistle blew.

HOUSING: UNCROWDED AND IMPROVING

Unlike mining companies elsewhere across southern Africa, Diamang elected not to house its African workforce in compounds. In company officials' estimations, the stability fostered by male workers' ability to remain with family members on the mines outweighed the security and control that the compound system supposedly delivered.[4] Diamang's housing units accommodated a maximum of eight single workers, as opposed to the dozens, or even hundreds of male laborers found in "overcrowded and diseased" compounds elsewhere in southern Africa.[5] Meanwhile, workers whose wives had accompanied them lived together with their spouses in smart, single-family structures.

To minimize labor and material costs, Diamang initially encouraged workers to self-construct these units. Over time, though, the company began building an increasing number of modern, or "permanent," houses for both single and married workers. Yet "traditional" stick-and-mud units remained a part of Diamang's housing strategy until the very end of its operations, especially on newer, provisional mines.

Regardless of whether structures were "modern" or "traditional," encampment populations were almost always demographically mixed. Single workers regularly lived next door to families, producing a level of conjugal and gender diversity foreign to the compound system. In fact, Diamang rarely segregated its settlements along these or other potential fault lines, such as ethnicity—a practice that T. Dunbar Moodie argues did much to facilitate the endemic faction fighting on South Africa's mines.[6] For the African men, women, and children living in Diamang's encampments, integrated accommodations encouraged intra- and interhousehold and intergender relationships, in contrast to male-dominated compound relationships. When coupled with the general absence of cramped conditions, these living arrangements generally minimized tension, rather than generated or exacerbated it.

In the following section, I explore the construction of housing on Diamang's mines, the ways the company populated these units, and the implications for the encampment residents. I also examine the location of these settlements relative to both work sites and water sources and, therefore, the ways that these spatial features shaped the extended labor process in which employees and family members remained engaged following their shifts.

The first priority for both single and married employees on reaching mine encampments was shelter. Diamang initially required recruits arriving from Dundo to construct their own accommodations, with the company assuming greater responsibility over time for this task. Although Diamang regularly improved housing construction, it wasn't until the 1960s that virtually all new accommodations featured durable materials (i.e., cement, brick, aluminum, and zinc), rather than grass, sticks, mud, and reeds. Yet even into the 1970s the company continued to erect (some) "traditional" housing, especially in the western reaches of its concession, as Diamang hastily expanded its operations into that area.

When contratados first began arriving on the mines from Dundo, mine bosses afforded them time to build their own *pau-a-pique* (stick and reed) housing in designated areas before they started formal labor. This construction, or "acclimation," period lasted anywhere from a few days to a week, depending on the individual mine boss and how urgently the new arrivals were needed on the job.[7] In general, this arrangement worked well for both the company and incoming workers: it excused Diamang from having to allocate valuable manpower and materials to the construction of housing, while allowing recruits to erect homes according to their predilections.[8]

In adherence to colonial notions of order and organization, the only stipulation the company made regarding worker-constructed housing was that it be erected in straight lines relative to existing structures (figure 6.1).[9] For company officials, orderly encampments reflected an orderly ethos, as did the obverse. As such, the company archive is full of photographs of workers' settlements, which Diamang employed as propaganda in efforts to emphasize its "enlightening" mission.

By the late 1920s, the company had begun to construct housing for its African employees, though in a highly limited fashion. In practice, Diamang was obliged to allocate even less manpower than it might have otherwise, as accommodations were often inherited from workers who had already completed their contracts and returned home.[10] For newly arrived recruits, the prospect of moving directly into a preexisting structure was undoubtedly welcomed, though in these instances

FIGURE 6.1. Mining encampment, 1938. *Source: Science Museum of University of Coimbra.*

Diamang also reduced their "acclimation period" to a single day. In theory, company representatives were to meet arriving workers and assign them to existing adobe or pau-a-pique houses, which Angola's high commissioner at the time, Filomeno da Câmara, praised as "the best that he had seen in the colony."[11] However, it would take years before the majority of workers received their housing in this manner. In the meantime, the company granted most arriving laborers time to construct their own accommodations.

Throughout the 1930s, Diamang officials vigorously debated the benefits of the compound system that so many of its engineers had personally witnessed during stints on South African mines. Ultimately, however, Diamang's interest in having families settle on its mines, as well as the perceived stability that this scenario provided, outweighed the allure of the compound system. Although a "closed" compound system would seem intuitive given the presence of conscripted laborers in the company's encampments, the paternalistic nature of Diamang's relationship with its labor force essentially obviated it.[12] With both absenteeism and desertion largely dissipating by the middle of the 1930s and diamond theft still not a grave concern, company officials determined that it was not worth sacrificing the existing stability that prevailed in the mine encampments.[13]

In 1936, Diamang signaled its commitment to improving accommodations for its African labor force by creating the Spamoi division, which had as part of its charge the construction and maintenance of housing in company encampments. As such, by the 1940s and into the 1950s, Diamang officials were echoing High Commissioner Câmara, declaring contratados' villages to be "models" for the rest of the colony. A 1941 company report described these settlements in the following manner: "Always situated next to roads, surrounded by hedges . . . each has a sentinela charged with looking after the collection of debris, the cleanliness of the houses, etc. . . . They all have enclosed spaces for the drying and grinding of manioc."[14] Similarly, a company official's letter to Diamang's director, Ernesto de Vilhena, in 1943 boasted that "it does not seem necessary to speak of the villages of our contracted workers, whose excellent state of cleanliness and appearance are already well known by your excellence, villages that can be considered as models, and are certainly without par in the colony."[15]

Beyond improving accommodations as part of its broader, strategic paternalism, Diamang also had fiscal motivations. Over time, company officials comprehended that due to the ongoing maintenance that "temporary" structures required, minimizing their number would be (more) cost-effective. Consequently, the 1960s witnessed an explosion of "durable" housing, composed of cement blocks or adobe bricks and featuring metal roofs and cement floors. During the decade, Diamang introduced men's and women's bathrooms, as well as washbasins for laundry at select encampments. The company also standardized integrated kitchens for all new housing and began upgrading old(er) houses, retrofitting them with cement floors or metal roofs, or both. Lina Machamba, who traveled to the mines with her husband, confirmed these improvements. "In the early 1970s, when we arrived for a second stint, this time on the new Chilumbuka mine, we received a nice brick house with a cement floor."[16] A report from this period outlines the features of the most advanced housing Diamang was offering at this time:

> In the most recently constructed villages, sanitary installations for both sexes were built with squat toilets and showers with running water. The supply of water, in some cases, is . . . delivered by an electric or gasoline pump. The water is piped to a common washroom, composed of twelve tanks for washing clothes in a covered building. . . . Also, there is electricity running to these so the exterior is illuminated. All the houses possess metal

beds for couples, and the dorms have, for single men, individual beds or bunk beds with two beds. . . . The kitchens, as a rule, are part of the annex of the house; they are ample and are connected to it by a closed yard.[17]

Figure 6.2 features an example of living quarters from this era. Again, the image is intended to demonstrate the orderly nature of Diamang's operations, but would also have been circulated to highlight the broader Portuguese "civilizing mission" in the face of the amplified international scrutiny that marked this period.

Despite increasing company involvement and overall improvements in workers' accommodations, even into the 1960s and 1970s Diamang was still using stick-and-reed housing. Thus, despite the good accommodations that most employees received following the initial decades of Diamang's operations, some were still constructing their own housing in virtually the same manner that African employees had been doing for years. For example, João Paulo Sueno stated, "I arrived here [Cafunfo] in 1973. Once I arrived, though, we still had to build our own houses! We had approximately a month to finish all the housing for the encampment, so we had to work together to do this. It was very close to both the river and the mine. When we finished, we went off to the mine to work."[18] Cafololo Muamuiombo confirmed that even when the company constructed workers' housing, it could still be

FIGURE 6.2. Housing for contratados on Fucaúma mine, 1961. *Source: Science Museum of University of Coimbra.*

"temporary" in nature, even into the 1960s. "When I arrived at the mines in 1963, there were houses that had already been built by the company [Spamoi]. They were made of adobe bricks and grass with no cement floors, and they were not painted. There were no latrines or bathrooms; we just went in the grass."[19] Although the company intended for some of these structures to be used multiple times, in most cases this housing was, indeed, only temporary—a stopgap measure until Diamang could replace them with more "durable" structures.

THE POPULATION OF COMPANY HOUSING

Just as the physical composition of workers' housing shaped their experiences away from the mines, so too did the ways Diamang populated these accommodations. By the end of the 1930s, the enterprise had settled on stand-alone houses with one, two, or three rooms, or in company lexicon: singles, doubles (the most common), and triples, with each room intended for three workers or one worker and his family. Even when Diamang introduced dormitory-style housing in the 1960s, these structures served fewer than a dozen, rather than hundreds, of single men, and featured divided rooms for privacy. In this sense, even when temporary overcrowding, due to influxes of workers, seasonal construction limitations, or the opening and closing of a mine, beset encampments, accommodations on Diamang's mines—whether company- or worker-constructed—were still far superior to compound housing. J. K. McNamara's contention that "violence on South Africa's mines is attributable . . . to a lack of personal privacy for workers crammed into living quarters" suggests that the general absence of cramped conditions at Diamang greatly contributed to the low levels of tension within its mine encampments.[20]

Although company officials occasionally flouted colonial housing capacity regulations, mindful of Diamang's labor attraction and retention objectives, these contraventions were hardly egregious and in no way reproduced the objectionable conditions of compound life. Even in Diamang's infancy, it generally lamented rather than ignored housing codes. For example, H. T. Dickinson, a Diamang engineer based in South Africa, wrote to Brandão de Mello, a Diamang official, in 1929 to "express my regret that . . . colonial legislation in Angola only allows six workers per compartment or housing unit. In South Africa, where I am, it's unlimited, but de facto, forty-eight."[21] Further, after a 1936 amendment to the Colonial Labor Code stipulating that "when workers had their families with them, the couple was entitled to their own room and children older than seven must have their own room," the diamond enterprise largely complied.[22]

By the early 1960s, the company reported that the average room on its mines housed one married couple or roughly 1.5 single men—well within acceptable limits of comfort—but that overcapacity continued to be a problem in certain encampments.[23] Informants who worked at Diamang during the 1960s and 1970s corroborated this admission, indicating that overcrowding was still an occasional problem, though most indicated that it was simply not an issue.[24]

As Diamang expanded, or at least varied, its contratado recruitment zones over time and began to provide accommodations for small, and then larger, numbers of voluntários by the end of the 1940s, encampments became increasingly diverse. Further muddling company settlements were blends of experienced contratados and new arrivals. Yet for all of the variety that Diamang's housing policies produced, this miscellany generated little tension. Testimony from Joaquim António Issuamo, who was on the mines from 1959 through Angolan independence, speaks to the multiplicity, the lack of segregation, *and* the relative amiability that the company's mine encampments featured:

> Contratados were from all different areas, so there were a number of ethnicities and languages in the same encampments because the company housed contratados from all over in the same encampments. So, of course, there were some conflicts along ethnic lines; of course, not all conflicts were ethnic-based, but at the end of the day, the company did not really care as long as the workers showed up for work. . . . There was a total mixture of ethnicities in mine housing, even sharing the same room.[25]

Testimony from Francisco Xamucuco, who worked for Diamang from 1958 to 1961, suggests that the company's apparent indifference was, in fact, a type of applied pragmatism. "There were no assignments based on ethnicity within the mine encampment; it was all practical—when there was an empty room it was filled, regardless of the ethnic group. . . . So, it was a huge mix."[26] Existing workers and newly arrived recruits represented another potential divide, but even these two groups interacted amiably in company mine encampments.[27] As Joaquim António Issuamo explained, "There was really no [obligatory] respect in the encampments for those who had worked longer with the company. . . . There was no power difference or animosity between workers with more, or less, experience."[28] Ultimately, the general lack of cramped conditions helped foster an environment in which demographic diversity produced relationships that transcended—rather than affirmed—real or imagined divides.

Every day except Sunday, workers tramped from their homes to their respective work sites, and then back again after their shifts ended. Commutes for both contratados and voluntários at times surpassed a dozen kilometers and could be dangerous if they included river crossings or other hazards.[29] The company generally aimed to position encampments within two kilometers of mines, but as mines were exhausted and operational areas drifted up and down river and creek beds—while settlements largely remained stationary—this objective often became logistically unfeasible. Consequently, many workers faced long commutes back home that cut into the time they had to prepare dinner. Immediately following an already taxing day at work, this trek constituted another challenge of the labor process at Diamang.

The earliest account of laborers' daily commutes comes from Lute J. Parkinson, who described a tragic incident from 1928 in which "the Chiumbe ferry sank and the master mechanic . . . lost his life. A large load of natives was also aboard, as well as two trucks. The overload made the gunwale too low. . . . The mechanic had jumped into the river in midstream and was swept away. . . . Days later his body was found miles downstream, badly mutilated by crocodiles."[30] Although Parkinson makes no further mention of "the natives," presumably some of them also lost their lives. Going forward, company records reveal that Diamang officials primarily concerned themselves with workers' commutes when mine encampments were located beyond the two-kilometer mark.

To remind Diamang of its reciprocal responsibilities, groups of male and female workers occasionally protested to company officials about commute distances. In a case from 1970, for example, a group of approximately forty-five men and women employed at Cabuaquece 2 mine complained about the distance of their daily traverse. Correia Oliveira, the mine inspector, filed the following report:

> About twenty women approached us and declared that they do not want to continue to come to work [each day], seeing

that they are housed in the village of Cabuaquece 2 mine and they have to come from there, which is far, maybe eight kilometers. . . . Consequently, they cannot provide the necessary assistance to their husbands and children, and they get very fatigued as a result of these daily journeys. On the mines, the husbands of some of these women also complained and asked to relocate closer to the mine, as where they currently are is very far to travel to work each day.[31]

Unfortunately, it is unclear what ultimately transpired, but company records indicate that Oliveira took the complaint quite seriously, and, thus, Diamang officials most likely acknowledged these employees' requests and acted accordingly.

Informants' testimony helps fill in the otherwise fragmentary written record regarding workers' commutes. Although most former employees recalled short walks to work each day, others cited distances that, at times, far exceeded the two-kilometer guideline; they spoke of assembling each morning to make the commute in groups. Mualesso Gaston, who started with Diamang in 1956, remembered: "It was ten kilometers to the mine from the encampment. We had to cross a river. We left the village at 4:00 a.m. in order to get to work by the 6:00 a.m. starting hour. . . . Different groups of guys would walk together every day and then split up when we got closer—some would go to the lavaria, some to the mine, etc. If you arrived late at all, even two minutes, you would be marked absent. You might as well return home after this!"[32] Itela Joaquim, who started working with Diamang in 1961, explained how distances could creep upward as time passed. "I was sent to Cula mine. There were already houses built there. Our encampment was three kilometers from the Cula River but one hour to the mine. The reason these encampments were so far away from the mines is that, at one time they were not, but as the mines kept closing and where the company was currently mining kept moving, the encampments got farther away and the company did not build new ones."[33]

LOCATION: ACCESS TO WATER

Until Diamang began providing running water to select mine encampments in the 1960s, workers and family members had to retrieve potable water from nearby rivers and streams. Beyond using water for hydration and hygienic purposes, it also had to be collected to prepare (boil) evening rations. Contratados' wives or children attended to this task before their return from the mines, but for unaccompanied men this portion of the extended labor process constituted an onerous responsibility after an already long day at work. The level of difficulty associated with retrieving water depended on the distance to the nearest river, stream, or creek from the encampment. And even after the company began supplying water to select settlements, it provided this amenity to only a handful of them. As such, manual retrieval of fresh water remained a daily feature of life for most African employees until the very end of Diamang's operations.[34]

For residents of encampments without an immediate fresh water supply, sources were located anywhere from "very close" to two kilometers away. Sacabela Sacahiavo's account of his encampment's proximity to the closest water source during the early 1950s was fairly common. "The potable water source was two kilometers away, at the river. Women and children would go there to wash clothes and to fetch water and bring it back to the encampment. . . . But these activities were only for their own benefit and for that of their husbands. Single men had to go to the river each day after work to get their own water and bring it back."[35] Similarly, Martam Camanda, who began at Diamang in 1969, declared that "it was only about one to one-and-a-half kilometers to the potable water supply from the encampment. Women brought water back during the day. But it was very difficult for (single) men to do this after they returned home from work."[36] To spread the burden around, Paulo Chingueji explained that he and a group of five other single workers "assigned a different person to retrieve water for the group each day."[37] Regardless, even this strategic arrangement remained inferior to the one enjoyed by workers whose wives had accompanied them.

In the 1960s, Diamang began furnishing running water to a small number of encampments. Cafololo Muamuiombo, a former employee, was a beneficiary of this initiative: "For drinking water, the company had built a lagoon and then something to pump the water up to the village. It was not inside the encampment but was only about two meters outside it. . . . We still washed our clothes, though, down in the lagoon."[38] For individuals like Muamuiombo and their wives, this relatively low-tech engineering solution greatly alleviated one of the more burdensome tasks that those who lived in Diamang's mine settlements faced on a daily basis.

FEEDING TIRED BODIES: NOURISHMENT
IN MINE ENCAMPMENTS

After procuring water, men and women next turned their attention to securing and preparing meals. Workers with accompanying family members again held a distinct advantage over their unmarried co-laborers, as wives and/or children handled this daily, time-consuming duty, enabling their husbands and fathers to rest on returning home from work. In this sense, family members played a socially reproductive role that all-male "brotherhoods" or other support groups often assumed on mines elsewhere in the region. Meanwhile, although the company had achieved food security by the 1930s, periodic shortages still punctuated Diamang's otherwise steady distribution.[39] In response, encampment residents lodged complaints but also traded with neighboring communities to augment or diversify food supplies. In the following section, I examine how employees met their food requirements over time, highlighting the roles that men and, especially, women played in this essential task.

COMPANY-ISSUED RATIONS

During Diamang's early years, the company struggled to furnish encampment residents with adequate evening and Sunday rations, which consisted of *fuba* (ground manioc), meat, salt, and palm oil. The meat and salt provisions were both small and sporadic, as these items were scarce, and white employees enjoyed first rights to any available supplies. In 1925, Eugenio Salles Lane, a Diamang engineer, explained the food-related challenges the company was facing: "The weekly ration is distributed to indigenous workers . . . in variable quantities consonant with the stocks at the company's disposal, which are not as abundant as the company hopes that they would be, owing . . . to the difficulty of obtaining the necessary items."[40] Food distributed to workers could also be of questionable quality, such as in 1927 when Diamang included sugar in Africans' rations from a particular shipment that was "unfit for whites to consume."[41]

In addition to engendering undernourishment among its male labor force, food shortages also affected the very women and children whom Diamang wanted so badly to attract to its mines. For example, while men were receiving 7,000 grams of fuba per week in 1927, wives were allotted half of that, along with 800 grams of dried fish and 300 grams of peanuts or palm oil. At this time, these women were theoretically supposed to be augmenting their family's food provisions with items grown in small plots adjacent to their homes using seeds provided by Diamang. However, due to inconsistent weather, limited time to allocate to this endeavor, and the lag between planting and harvesting, it is unlikely that many families benefited from this company initiative.

Only as a result of Diamang's implementation of ambitious food production and regional purchasing schemes—all part of the company's pragmatic/paternalistic measures—did the company achieve food security beginning in the 1930s.[42] Diamang also diversified its provisions around this time. In 1927, for example, the company distributed 7,000 grams of fuba weekly to workers, but a decade later it reduced this figure to 5,250, substituting additional meat, fish, and peanut rations and introducing 700 grams of beans or rice (depending on supplies).[43]

Former employees attested to the general reliability and reasonable diversity of company rations over time. For example, Deque, a former employee who arrived on the mines in 1956, described the weekly distribution of food as follows: "We would collect our rations at different stands . . . for example, one for fuba, one for dried fish, etc. . . . at the

FIGURE 6.3. A rare image of laborers preparing dinner in a mine encampment, 1938. *Source: Science Museum of University of Coimbra.*

end of work on Saturday and then take them home for the week. Each person had a number, and when he was called he could then begin . . . collecting the rations. I always got a complete ration; there was never anything missing. I was never shorted anything."[44]

Into the 1970s, the food security that Diamang had achieved decades earlier ensured that ration quantities and daily caloric consumption had changed little over the years. For example, in 1940 the average caloric intake in grams was 4,145 and in 1962, 4,152, while in 1971 the company was distributing weekly rations as shown in table 6.1.

Considering that the rations distributed to married workers were meant for two, but that the company intended the husband to consume as much as a single worker, women obviously received something shy of a "full male ration" (table 6.1). Yet compared with the food insecurity that workers had suffered through decades earlier, this feature of Diamang's food allocation practices was largely tolerable and, in any event, constituted a significant improvement on earlier scenarios.

ENCAMPMENT RESIDENTS' SOLUTIONS TO FOOD SHORTAGES

When intermittent food shortages did occur, Africans living in mine encampments creatively responded. Throughout Diamang's history, these residents issued rations-related complaints intended to prompt the company to take remedial action and, far removed from any urban centers, also traded with neighboring communities to augment and/or diversify their food supplies. Although company representatives were typically responsive to workers' grievances, they consistently forbade workers and family members from using their company-issued

---

TABLE 6.1

Weekly rations for married and single workers (in grams)

Item	Married workers	Single workers
Fuba	8400	4900
Dried fish	2100	1400
Peanuts	560	350
Beans or rice	1050	700
Salt	140	70
Palm oil	140	70

Sources: Diamang, Breve notícia 33; Fernando Duarte, "Nota de informação no. 155/71, Visita à Mina Gungo 2" (December 24, 1971), 2, MAUC, Folder 86A7 31°; A. Pinto Ferreira, "Notas sobre a alimentação dada aos seus trabalhadores" (April 13, 1945), 2, MAUC, Folder 86D 4°.

---

rations or garments as truck for food. As Barun Mitra has commented, "When trade is outlawed, only outlaws trade."[45] Accordingly, Diamang instructed sentinelas to monitor this otherwise harmless commerce and to punish those who engaged in it. Given the open nature of mine settlements and frequent contact between encampment residents and neighboring communities, though, this charge was difficult at best, and, in practice, company officials displayed considerable apathy toward the enforcement of this directive.

Complaining for More

Just as encampment residents complained about other disagreeable aspects of life at Diamang, they also periodically articulated rations-related grievances as part of the wider "culture of complaints" they cultivated at Diamang. The first recorded account comes from 1962 and chronicles a development in which workers' wives had begun to openly disapprove of the quality of their rations, spurning company offerings, and demanding improvements.[46] A company report describes this process and the eventual solution: "The women . . . would not show up to receive the rations that were destined for them and their children. . . . They refused . . . to receive their rations, especially the manioc, which had been placed at their exclusive disposal."[47] Diamang subsequently rectified the situation by conceding that "women can now go to individually select the manioc they desire, and we are also now treating the *bombó*, obtaining a fuba 'to their liking.'"[48] Shortly thereafter, in another incident concerning rations, a contratado named Muaguite objected that encampment women had received fuba only four or five times during a span of nineteen months.[49] It is unclear how officials reacted to Muaguite's complaint, though the level of concern expressed in ensuing company letters and internal documents suggests that Diamang most likely attempted to redress this shortage.

Commercial Contact with Neighboring Communities

Although encampment residents had surely been trading for food supplies with neighboring villagers for decades, not until the 1960s do written and oral sources offer a glimpse into the vibrancy of this commerce.[50] In an incident from 1963, which precipitated a spate of correspondence between Dundo and Lisbon, the company uncovered a clandestine market near Chambuáge mine at which workers, their wives, and local residents would meet each Sunday to trade

company-issued dried fish rations for manioc and sweet potatoes. Filipe Saucauenhe, a former contratado who began at Diamang in the
1960s, confirmed that these types of commercial relations persisted.
"I had a friend who lived in a nearby village who also worked on the
mine. People from these villages often worked on the mines [as vol-
untários]. . . . We could visit these villages. We traded for, or bought
from neighboring people, mangos, sweet potatoes, and fish—but
only food. . . . We traded our company-issued oil and salt in exchange
for food. This was illegal."[51] Contratados and family members also
engaged in this trade to diversify their food supplies. João Paulo
Sueno explained that "my family had lots of food because all three of
us [he and his parents] were working. We used to go to neighboring
villages to exchange fuba and peanuts for chickens because Diamang
only ever gave us dried fish!"[52] Not surprisingly, friendships often blossomed from these commercial exchanges.

SOCIAL RELATIONS WITHIN MINE ENCAMPMENTS

Workers' and family members' relationships with company officials,
sentinelas, and fellow encampment residents, as well as with members of neighboring communities, played a significant role in shaping
their lives away from the mines. In these typically relaxed settings, men,
women, and children actively cultivated an array of relationships that
transcended ethnic, linguistic, marital, and tenurial divisions, which,
in turn, collectively helped them improve their lives. The uncrowded
conditions, ethnic homogeneity, and the stabilizing presence of women
in Diamang's encampments combined to create a unique atmosphere
that generated little of the animosity that pervaded the cramped, male-
dominated compounds elsewhere in southern Africa. Even company
officials and sentinelas typically acted as mediators in mine encampments rather than as hostile overseers. and female residents only occasionally experienced gender-based aggression. Even when women did
figure centrally in male laborers' disputes, these men rarely entered
into the sorts of violent confrontations with co-workers that profoundly
unsettled mining compounds elsewhere on the continent. The following section explores the shifting web of largely amicable social relations
in which laborers and their family members daily engaged away from
the work site, including with company representatives, fellow encampment residents, and members of neighboring communities.

After men and women left their respective work sites each day and returned to mine encampments, they typically came under the vigilance of African sentinelas, who also resided in these settlements. In the absence of sobas, whom the company kept away from the mines in a deliberate attempt to displace them, sentinelas arbitrated disputes, encouraged residents to practice restraint—such as by restricting parties and drinking to Saturday nights—and punished any transgressors. Conversely, European officials visited encampments only to intervene when residents were unable to resolve protracted interpersonal issues. Despite their different levels of involvement in encampment life, both sentinelas and white Diamang officials generally interacted well with mine encampment residents, assailing only those whom they deemed to be "troublemakers."

From early on in Diamang's history, white company representatives were involved in settling intra-encampment disputes that residents either brought to their attention or that they identified during periodic visits to the settlements. Parkinson's description of an episode from the 1920s suggests, however, that these mediations were not always satisfactory. "My Belgian friend . . . once tried to decide a native *ndaba*. These palavers or quarrels might go on for hours and the European, in attempting to settle everything fairly, might fail to satisfy anybody. On this occasion, in boredom or disgust, Bob gave a careless verdict only to discover that both parties were badly dissatisfied with his decision."[53] As time passed, white company officials displayed increased levels of familiarity with local culture—including conflict-resolution methods—and by 1949 Diamang was compiling statistics concerning their mediative efficacy (table 6.2). These figures suggest that over time these Diamang representatives improved their ability to successfully resolve disagreements.

When intervention by white company officials was either undesirable or impracticable, encampment residents often enlisted sentinelas to act as "big men" in the resolution of otherwise intractable disputes. Testimony from Dinis dos Santos Muriandambo, who was with Diamang from 1969 to 1975, confirms that these African overseers played key roles in settling quarrels. "Our sentinela . . . would summon the *mais velhos* [the eldest] in the encampment to resolve problems in the same manner a soba would, in a traditional African manner."[54] Whether company officials or sentinelas played this arbitrative role,

TABLE 6.2

Number of disputes handled by Diamang and the percentage resolved

	1949	1951	1952	1954	1959	1960
Total number	n/a	n/a	187	220	100	64
% resolved	65%	68%	n/a	92%	n/a	n/a

Sources: *Spamoi relatório anual de 1949, mão-de-obra indígena* (January 27, 1950), 44, MAUC, Folder 86D, 2 5°; *Spamoi relatório anual de 1951* (February 1, 1952), 38, MAUC, Folder 86D, 2 6°; *Spamoi relatório anual de 1952* (February 14, 1953), 36, MAUC, Folder 86D, 2 6°; *Spamoi relatório anual de 1954* (February 12, 1955), 38, MAUC, Folder 86D, 2 7°; *Spamoi relatório anual de 1959* (February 16, 1960), 64, MAUC, Folder 86D, 2 12°; *Spamoi relatório anual de 1960* (March 21, 1961), 32, MAUC, Folder 86D, 2 13°.

their involvement reflects Diamang's calculated supplantation of sobas—a fundamental component of its broader paternalistic strategy.

RELATIONS AMONG ENCAMPMENT RESIDENTS:
FRIENDSHIPS AND FRACASES

Social relations within mine encampments were largely concordant. When disturbances did occur, they almost always involved male employees and two proverbial issues: women and material possessions, often fueled by alcohol. However, examples of tolerance or indifference far outnumbered instances of hostility or aggression. Men, women, and children actively cultivated friendships and enjoyed the attendant opportunities to both commiserate and cooperate.[55] In practice, this undertaking was often predicated on personal compatibility and generally transcended, rather than respected, a variety of potential divisions.

Intra-Encampment Tension and Disputes

Written accounts of intra-encampment acrimony are not available until the late 1940s, the point at which Diamang had started recording the types and quantity of the disputes that reached company authorities. Yet the range of root causes presumably also reflects the impetuses for disagreements solved *without* company officials' awareness, or at least without their intervention. The figures are revealing in that they demonstrate that material issues constituted a significant percentage of the overall disputes, although informants also suggested that women featured in more quarrels than the figures indicate (table 6.3). As Dinis dos Santos Muriandambo disclosed, "There were problems when a worker owed . . . money, when people were drunk, or when disputes over women arose."[56]

TABLE 6.3

Nature of disputes handled by Spamoi

	1949	1952	1954
Debts	40%	40%	46%
Thefts and fraud	24	20	21
Aggression/assault	15	12	3
Conjugal issues	9	10	9
Diverse/unfounded	12	18	19

Sources: Spamoi relatório anual de 1949, mão-de-obra indígena (January 27, 1950), 44, MAUC, Folder 86D, 2 5°; Spamoi relatório anual de 1952 (February 14, 1953), 36, MAUC—Folder 86D, 2 6°; J. Robalo, SPAMOI relatório anual de 1954 (February 12, 1955), 38, MAUC, Folder 86D, 2 7°.

Although salacious advances appear to have been rare, and most female informants declared that they never experienced problems of this nature, women did occasionally receive unwanted attention from aggressive male co-residents. As Charles van Onselen has pointed out, many mine workers had trouble protecting wives in these new settings, within which the rules of village life often seemed not to apply.[57] Muriandambo, a former employee, explained that "problems arose when aggressive younger men went after the wives of older men. Always the younger ones wanted the women of the older ones."[58] Paulino, a former contratado, admitted that he was one of these suitors: "For the first three or four weeks I stayed out of trouble. After a month, it was natural that I would start getting in trouble. Fights were usually over someone owing someone else money, drinking, or having an affair with a woman."[59] Paulo Chingueji, a former employee, indicated that these altercations could, in fact, have severe consequences. "Sometimes wives caused problems between men; the women provoked these. Sometimes fights broke out over women, which would result in the parties losing their possessions . . . and going to jail."[60] Most women, however, neither instigated nor were enmeshed in these types of incidents, successfully avoiding this sort of attention. Reflective of the social professionalism that, over time, predominated within the African labor force, Txipanda Armando, a former employee, stated, "The women in the encampment . . . did not cause problems; if you were there to work, then you did not cause any problems."[61]

Tolerance and Amiability

Uncramped conditions, ethnic and linguistic homogeneity, and the stability that workers' wives provided rendered Diamang's encampments sites of tolerance, respect, and amiability. Even settlements that were ethnically diverse generated little hostility and were marked mainly by a camaraderie that transcended ethnic and other distinctions. Those recruits who had traveled together to the mines and managed to stay together through the deployment process enjoyed relationships that predated their time at Diamang. Meanwhile, women just as actively forged and deepened relationships while on the mines, primarily with their husbands, but also with other female residents. In fact, many married couples declared that they strengthened their existing relationships as a result of their shared Diamang experiences. Like their parents, children also made friends in the encampments. Collectively, this complex web of relationships helped residents negotiate these foreign environments.

Although ethnic tension was a hallmark of mining accommodations across southern Africa during Diamang's operational era, residents of its encampments appear to have paid ethnic or even linguistic variations little heed. Figure 6.4 was almost certainly intended for propaganda purposes, but it also captures the amiability that most informants claimed pervaded the encampments. José Silva, who began with the company in 1950, recalled that "I lived together with four to six others. . . . I lived and worked with other youths and lived in the encampment with them and all sorts of others from different ethnic groups . . . who became my friends."[62]

FIGURE 6.4. Gathering of workers in a mine encampment, 1939. *Source: Science Museum of University of Coimbra.*

Encampment residents also largely ignored linguistic divergences as they strategically cultivated relationships in these heterogeneous settings. In practice, this endeavor was made easier when languages were grammatically or otherwise similar, as Caiombo Jombe explained: "[In 1966–67] I lived with Chokwe, Kacaris, and Lundas in the encampment. . . . Ethnic groups were not separated in these villages. We were mixed among each other. In fact, even though language served as somewhat of a barrier, because the languages were often so close to one another we could understand each other."[63] For many employees, diversity of any kind was perceived as a social opportunity rather than an obstacle. Testimony from Mateus Nanto, an ethnic Chokwe at Diamang in the early 1960s, is characteristic. "I always lived with a friend from my village, and four others who were ethnic Bangala, with whom I became friends. All six of us [roommates] became good friends."[64]

Workers and family members in company encampments also seem to have disregarded marital status as a potential social impediment. When I asked Martam Camanda if he had single friends or if he, as a married worker living with his wife in an encampment in the 1970s, only congregated with other couples, he said, "I trusted both couples and single workers. I ate, drank [with], and went to the homes of single workers, too."[65] Likewise, many single laborers explained that they befriended married workers. Bernardo Montaubuleno, an unmarried worker, laconically remarked that "There was no division between workers with families and single men in the mine villages."[66]

Laborers also appear to have rejected segregating themselves according to levels of seniority, as most informants suggested that any social discrimination that may have existed along tenurial lines was fleeting. According to Paulino, who was at Diamang in the early 1960s, "After a short time, the newly arrived workers got along with the existing workers. Forging friendships . . . had everything to do with personality—the same way you always make friends."[67] Dinis dos Santos Muriandambo echoed Paulino's sentiments, recalling that "when I arrived in the encampment [in 1969] . . . there were already people living there. We all lived together, mixed up between the people that had already been there and the newcomers. . . . In the encampment, even though you lived among the existing residents, for about the first three weeks you were really living in separate worlds. But this barrier was dissolved within about three weeks."[68]

The handful of workers who managed to stay together from initial recruitment to deployment were not obliged to reach across ethnic, linguistic, or any other type of potential divide to forge relationships—even if they often did anyway. António Muiege was among those employees who were fortunate enough to have remained with fellow villagers, first as recruits and then as roommates within company encampments. "In 1960, I went to Camutala mine. . . . I lived in the same encampment as those who worked on the mines, and in the same house with guys who came from my village. We all had different jobs . . . but we would eat and cook together every night, so we basically remained best friends."[69] Of course, this type of arrangement was impossible to engineer given the arbitrary manner in which company officials assigned incoming recruits to individual mines, but Muiege's experience was certainly not unique.

Female informants also emphasized the friendships that they had established, insisting that these relationships had not only helped them survive the challenging environment but even to enjoy it.[70] Mulevana Camachele, who lived on Maludi-Page mine in the early 1960s, remembered warmheartedly that "the majority of my friends lived in the encampment. I had one friend in the encampment with whom I had a strong, strong friendship—from the heart [she motioned at this point, drawing a fist up to her heart]. We were never separated. I met her in the encampment."[71] As with their male counterparts, women's relations also transcended potential divides. Mawassa Mwaninga, who lived on Maludi mine in the mid-1960s, described her experience forming a series of new friendships after the company had assigned friends from her village to different jobs: "Many women I encountered at Diamang had already worked for a while and had experience. These women only gave us work advice. They did not speak poorly of whites or of the company. . . . I made new friends with these women with whom I worked in the fields. My original friends were split up job-wise; one worked in the fields and the other on the mine. The white boss decided who worked where."[72]

Even as women were forging new friendships within mine encampments, the majority of their personal interactions were with their spouses. According to cultural norms, women assumed subservient roles within households, which were primarily intended to facilitate social reproduction. When I asked Felipe Leo Muatxissupa if his wife worked for the company, he replied, "No, only for her husband," referring to the custom of women in the encampments to cook, retrieve water, clean, and take care of any children present.[73] In this sense, men with wives at Diamang relied heavily on their spouses, while single men looked enviously upon those workers whose female partners had accompanied them. As testament to how extreme this form of conjugal dependency could become, Itela Joaquim bemoaned: "The only problem with having a wife there was that when she got sick you had to do all of the chores yourself!"[74]

Despite, or perhaps because of, husbands' profound reliance on their wives, many female informants explained that they deepened their matrimonial relationships while living in the encampments. Janette Pedro recalled the daily treks to the mine and back as bonding experiences, rather than as mere bookends to otherwise taxing days spent laboring. "I walked to and from work with my husband. We left at the same time in the morning to go to work, and then I would wait for him to finish work—I would be done earlier—so that we could walk back together. We were in love!"[75] Testimony offered by João Paulo Sueno, who was a minor in his parents' care on the mines in the 1960s, features similar expressions of conjugal and familial felicity, also related to the daily commute: "My mother, father, and I all worked on the same mine. We walked together to work in the morning and home at night together. We were very happy."[76]

In addition to deepening relationships with their parents, minors in the company encampments also enjoyed interacting with other children. Although these youths are invisible in company and colonial records, many informants remembered their childhood experiences in the settlements warmly. Most children helped gather water and prepare meals, but like their parents, they also found ample time to socialize. According to João Saluembe, who first arrived at Diamang in 1950, when he was nine years old, "I only saw my dad at night because we worked in different places. We cooked together at night for each other. . . . I lived and worked with other youths and lived in the encampment with them. . . . They became my friends. Kids really were not excluded from much. It was very open, and we all . . . went in and out of each other's houses."[77] Bernardo Montaubuleno, who first arrived on N'Zargi mine as a child in 1953, echoed Saluembe's testimony, recalling that "kids on the mines all ate and played with one another and in each other's houses. . . . Later on, I met some of them in Saurimo, and everyone still received each other well."[78] In fact, in an

FIGURE 6.5. Children playing in Caingági mine encampment, 1966. *Source: Science Museum of University of Coimbra.*

attempt to entice men to bring their families with them to the mines, beginning in the 1960s Diamang installed recreational equipment in some of its newest encampments (figure 6.5).[79]

SOCIAL RELATIONS BEYOND THE ENCAMPMENT PERIMETER

Beyond interacting with members of neighboring communities to augment or diversify food supplies, mine encampment residents also forged friendships with the villagers. In rare cases, amorous relationships blossomed from these encounters, but interaction of this sort could, at other times, prompt aggressive action over local women. Several factors determined the exact nature of the contact between local communities and mining encampment residents, including proximity, company representatives' level of tolerance, and the individual personalities involved. It remains unclear whether commercial interaction facilitated these cordial relationships or the other way around, but they were often just as valuable as intra-encampment personal interactions in helping residents cope with daily life at Diamang.

Former contratado encampment residents and neighboring villagers who worked as voluntários both offered testimony that highlights the depth and importance of these relationships. According to Deque, a former voluntário, relations could be quite profound. "My village and the nearby contratado village had parties together. They would visit us as friends. Contratados . . . were invited on Sundays . . . to my village and to other villages to drink."[80] When I inquired if language barriers were obstructive or ever caused problems between contratados and residents from his village, he tersely replied: "No, we were all Angolans." Women, too, forged meaningful relationships. According to Mulevana Camachele, "We traded with neighboring villages — things that we had run out of or things we didn't have. I developed a strong friendship with a woman named Elena in a neighboring village. She helped me with salt, gathering firewood, etc."[81] Although certain company officials harbored misgivings concerning this type of interaction, most were indifferent. According to João Paulo Sueno, a former contratado, "Sometimes people from the encampment would go to the neighboring villages on Saturday nights to drink and celebrate. The company knew about this but kind of looked the other way. It was not permitted, but it was tolerated."[82]

Other times, devoted relationships grew out of contact between male encampment residents and females from local communities. Caiombo Jombe explained that "some contratados had girlfriends from neighboring villages not associated with the mines. There were many relationships between contratados and non-contratados."[83] Moreover, Deque explained that matrimony could even result. "Marriage was possible between those in the contratado encampment and someone from a village. In fact, someone from my village married a contratado, who subsequently took her back with him [to his home village] after his contract had ended. The company did not mind these types of developments."[84] For an encampment resident who was missing home, the process of forming a new family undoubtedly helped counter this sense of absence.

Just as men's competition over women within mining encampments periodically generated interpersonal problems, some informants indicated that contratados' pursuance of women from neighboring communities could also trigger disputes. Paulino, who started at Diamang in the early 1960s, outlined this type of scenario: "In many cases, trouble would occur when contratados living in the encampments would go to nearby local villages with fish, or other items, trying to attract women. Their efforts would then create tension or even cause fights between local villagers and these contratados."[85] Filipe Saucauenhe explained that irate men from neighboring villages might even kill the contratado suitor if the amorous relationship "violated a tradition."[86] Yet for most of the men and women from either mine encampments or neighboring communities, relations were largely affable, even when they didn't lead to marriage.

COMPANY-SPONSORED RECREATIONAL ACTIVITIES

Over time, workers and family members living in Diamang's mine encampments had opportunities to develop and expand relationships via a range of company-sponsored extracurricular activities, including film screenings, athletic competitions, and the monumental *Grande Festa* (Grand Festival) celebration.[87] Via the provision of these activities, Diamang officials were attempting to defuse potentially explosive occupational stress by diverting attention away from the daily work regimen. These recreational offerings also constitute manifestations of company officials' increasingly paternalistic mind-set from the 1930s forward. By providing this range of popular and beneficial services for its African labor force, Diamang hoped to further stabilize it.

THE GRANDE FESTA

In the 1940s, the company formalized the annual Grande Festa, which had evolved from a series of earlier celebrations. The festival, which was held each August, comprised a succession of events, including an Olympics-style sports tournament in which African contestants competed for the "Commandant Ernesto de Vilhena Cup," and the "Best Village Competition," that is, the cleanest and "most hygienic" encampment (figure 6.6). The hurdlers' uniforms and patent exertion, the decorated grandstand, and the lined track all suggest the seriousness with which both the company and the participants took this competition. These events were followed by three to four days of dancing and other "traditional" performances by local artists, as well ceremonies during which Diamang publicly awarded and rewarded winners of the sports competitions, residents of the Best Village, and veteran employees.

According to the company, the Grande Festa was organized "for its importance as a normalizing element for . . . types of entertainment pernicious to the equilibrium of social discipline."[88] In particular, Diamang officials were particularly pleased with the way that the Festa advanced its "civilizing" and stabilization efforts in Lunda: Africans

FIGURE 6.6. Grande Festa, c. 1950

heightened their efforts to keep settlements clean in pursuit of the Best Village prize, while the company presented, and financially rewarded, experienced voluntary workers as paragons for others.[89]

In addition to rewarding long-standing employees during the Grande Festa, Diamang also honored "cooperative" sobas, understood by company officials to be those headmen who consistently furnished workers; encouraged the sale of local produce to the enterprise; and, in every other respect, caused the company no trouble At the 1957 Grande Festa, Diamang hosted 525 sobas, who were themselves accompanied by a further 405 women (presumably wives) and 285 children. While in attendance, the company fed the sobas and also distributed wine and cigarettes, and they and their families stayed in specially renovated housing. Yet despite Diamang's efforts to publicly thank the sobas for their contributions over the past year and to ensure continued compliance going forward, not all of them appreciated the gesture. Some six years earlier, in 1951, Diamang reported that some of the sobas in attendance were upset that the company failed to offer them any money, while openly presenting cash rewards to veteran employees, whom these disempowered, traditional authorities still considered to be their "subjects." As one soba from Sombo, who had refused to attend the Festa that year, allegedly griped about past festivals, "I liked them a lot, I ate and I drank as much as I wanted. But money is what Diamang does not give me."[90] After being summarily displaced by the diamond company in Lunda and banned from the mines for all but four days a year, some sobas remained understandably disgruntled.

The majority of the company's African workers (and sobas), however, enjoyed the Grande Festa, even if it was small consolation for the energy that they had expended on behalf of the company over the course of the year. Informants remembered the series of events fondly, especially the dancing and feasts of oxen. As Fernando Meuaçefo, who worked for Diamang from 1967 to 1975, exclaimed, "I liked the Grande Festa very much! It lasted for four days, and people came from all around."[91] Irrespective of the company's calculated motives for the Festa, it represented a welcomed and enjoyable reprieve for workers and their family members and, thus, was greatly anticipated.

POPULAR CINEMA: EVENING FILM SCREENINGS

Unlike the gala, but only once-per-year, Grande Festa, film screenings and company-sponsored athletic events provided workers with more frequent recreational outlets and were just as popular. These activities were relatively inexpensive for the increasingly prosperous enterprise to offer and company officials considered them to be "wholesome" pursuits that were intended to build a type of recreational dependency and, ultimately, to maximize profits by helping to stabilize the mines.

In the 1950s, Diamang introduced highly popular "mobile cinemas" so that employees and their families could view films on select evenings during the dry season. The company's stated design was "to show films with a simple plot, meant to demonstrate the advantages of medical aid, proper hygiene, and steady work."[92] Although it is difficult to imagine how some of the films might have accomplished these ends, the company nevertheless carefully selected each of them. Films were of varied genres, though were quite often documentaries highlighting some aspect of life in Portugal that, if not made under the auspices of the metropolitan regime, had already been censored by it for any potentially subversive content. The films typically featured locales either in Portugal or its overseas empire; they included *Imagens de Lisboa*, *Aspectos do Porto*, *Aveiro*, and *Macau*.[93] Other films, such as the 1952 screening of *Pais brancos—Missionários da Africa* (White fathers—missionaries of Africa), included scenes that illustrated positive interaction between Europeans and Africans, either within or outside of Angola. Similarly, the company's 1953 showing of *Bambuti*, documenting the life of Mbuti "pygmies" in the Congo, was intended to demonstrate how African employees at Diamang were advancing as a result of their regular contact with whites—in contrast to the "primitive" Bambuti shown in the film. Occasionally, films were meant to be strictly informational, such as the 1956 viewings of *Insectos que transmitem doenças* (Insects that transmit diseases) and *O que é a doença* (What is a disease?), though company officials noted that African viewers typically laughed at these offerings, apparently finding them humorous.[94] Finally, on rare occasions, the company screened films with little value outside of pure entertainment, including, for example, the 1966 showing of *Ben Hur no fossa dos leões* (Ben Hur in the lions' pit).

On days when the itinerant cinema was to arrive, the company placed flags in the encampments to signal the planned activity for that evening. A Diamang report from 1954, filed shortly after the cinema first appeared, declared that on encountering the flags, "The Africans love it and run around shouting 'cinema.'"[95] Sessions generally lasted an hour and a half, and films were typically prefaced with introductory commentary in local languages—primarily Chokwe—so that viewers could better understand what was to follow. Screenings were quite popular, averaging five hundred to six hundred people, and at times drawing over a thousand, and thus tens of thousands annually.[96] Given that attendance was voluntary, film screenings constituted an activity through which both Diamang and its African labor force actively stabilized the company's mines.

ATHLETIC ENDEAVORS

In addition to the athletic events featured at the annual Grande Festa, company officials felt that more frequent physical activity would enable workers to better enjoy the free time they had after their shifts ended each day and in anodyne fashion. Therefore, beginning in the 1940s, the company began introducing soccer balls into contratado encampments, and yo-yos, puzzles, and tops into the secluded Selection Station housing, all of which were reportedly well received by the men and boys residing within them.[97] Owing to the ongoing popularity of soccer, the company later organized teams, and, in 1970, officials attempted to increase workers' morale and allegiance to their respective mines by forming a rudimentary soccer league that pitted teams from individual installations against one another. The decision was justified as follows: "With the objectives of satisfying the workers, keeping them occupied outside of their service hours and, eventually, obtaining a greater attachment to the mine at which they work, it was decided to facilitate the practice of soccer for our labor force, foreseeing the realization of matches between teams representing the diverse mines, and eventually . . . an inter-mine championship."[98]

ENCAMPMENT ACTIVITIES AS CULTURAL CONNECTORS

Workers and family members also developed and expanded their assortment of social relations with fellow encampment residents, as well as with neighboring communities, through a range of their own activities. In addition to the company offerings outlined above, workers contented themselves with familiar pastimes, including drinking, dancing, singing, drumming, and reproducing cultural ceremonies. These pursuits were primarily social in nature, but they similarly helped workers relieve the tension that unfailingly built up during intense workweeks. Consequently, Diamang officials largely tolerated these "traditional" endeavors, even as these same officials were simultaneously busy promoting the range of company-sponsored leisure activities on offer. Only brewing alcohol, attempting to trade company-issued items for it, or listening to the nationalist movements' radio broadcasts elicited aggressive company responses, thereby ensuring that practitioners pursued these three activities clandestinely.

CELEBRATION: DRINKING, DANCING, SINGING, AND DRUMMING

Men and women residing in mine encampments initiated and cultivated friendships each day after work, but especially on Saturday evenings. On these nights, workers and family members gathered to dance, sing, drum, and, typically, drink. Elsewhere, Emmanuel Akyeampong has demonstrated the powerful roles that alcohol has played in facilitating community and identity.[99] At Diamang, challenging work conditions thrust workers together; alcohol subsequently helped them initiate and then deepen an array of resultant relationships, which emphasized collective, rather than individual, identity.

Company officials condoned these activities and even privately encouraged this revelry. To retain control over drinking, though, Diamang forbade residents to brew alcohol within the encampments or trade for it with neighboring residents, forcing men and women to pursue these activities covertly. Company officials also expected that workers would use Sunday to adequately recover and over time grew increasingly intolerant of any violence or truancy stemming from "excessive" Saturday night celebrations. According to João Paulo Sueno, "If drinking at all affected your work on Monday, then you would be beaten because the overseers knew who was drunk, or still hungover, or still drinking on Sunday."[100] Over time, however, workers' increasingly professional approach to employment rendered these occurrences extremely rare.

Dancing, Singing, and Drumming

Parkinson provides the first recorded insight into employees' dancing, singing, and drumming.[101] As a prospector, he enjoyed an intimate perspective on these activities, and his account of Saturday evening celebrations in the 1920s reflects this positionality:

> As a general rule, no work was done on Sundays. Beginning on Saturday night and continuing through Sunday evening, the drums would be rumbling constantly. As a result, the natives were rather bleary eyed when they presented themselves for work on Monday morning. Although the men usually took the leading part, men, women and children participated in the dancing, separately or together. The dances were usually rather suggestive, accompanied most often by one person singing and the entire chorus crashing in from time to time as they danced. . . . This sort of thing would go on for hours and hours until the poor white man lying in his camp was inclined to become somewhat impatient and irritated at such displays of sustained high spirits, but the wise man considered it good psychology and . . . encouraged them to keep up with their fun.[102]

Informants also expressed favorable opinions of this gamut of activities, indicating that dancing and drumming, in particular, were effective in "taking the edge off" and uniting the encampments after long work-weeks. Bernardo Montaubuleno, for example, recalled that "there was dancing on Saturday nights. Songo, Chokwe, etc. all danced together and got along. There was no trouble."[103]

Singing was also an integral component of these gatherings, though it was not necessarily limited to Saturday nights. This endeavor provided workers and family members with opportunities to both reinforce connections to their now-distant home villages and forge new friendships by learning songs from one another, at times in unfamiliar languages or dialects. Isabela Casombe, who was from western Lunda, indicated that during her time in the encampments in the 1960s, "I learned Chokwe songs at night, and there was a great mixture of people singing."[104] Testimony from Luciano Xacambala, who lived on Mussolegi mine in the early 1950s, confirmed that singing also unified camp residents, even when learning new languages was not required. "We sang in the camp at night. They were only songs about life, etc.—traditional songs about things. . . . We all had different songs and dances, but we sang and danced together, and each group attended the others' performances, dances, etc."[105] In this respect, singing represented a medium through which residents both initiated and deepened friendships.

Most informants affirmed that in addition to reconnecting with their homelands, they also occasionally used songs to disparage the Portuguese—though not necessarily Diamang—seeking fellowship through the articulation of pejorative or rejective lyrics.[106] Miudo Rafael, who started on the mines in 1962, recalled: "In our encampment, we would often sing songs at night—some nasty ones about the Portuguese. We had plenty of these."[107] He then shared a song with me related to the beatings and attendant commotion that occurred at colonial administrative outposts (postos):

> *Speak clearly. Speak clearly.*
> *Right at the posto there is always noise, always.*
> *I will go to the posto to see where my father went so that I will also know*
> *If there is any noise.*

António Batista indicated that the violent outbreak in 1961 of the Angolan War for Independence and the subsequent arrival of the PIDE (the Portuguese secret police) prompted residents to be more discreet when engaged in this activity. "People would sing songs about the Portuguese before the PIDE arrived, but then they became worried about informants, so they sang them much more quietly or not around certain people. Songs after 1961 really scared the Portuguese as they saw what had happened."[108] Even if voices had to be lowered on occasion, though, it is clear that singing was both an important and durable pursuit for members of Diamang's African workforce.

Alcohol: Access and Consumption

Alcohol consumption fueled the range of leisure activities associated with initiating, cultivating, and deepening friendships. Diamang officials adapted colonial regulations regarding African alcohol consumption for their own purposes, forgoing the most stringent restrictions and deeming that, though not wholly desirable, controlled drinking would ultimately be beneficial to its bottom line.[109] Away from the mines, male and female residents both legally and illegally procured alcohol by buying or trading for it, while small numbers of female brewers covertly manufactured it. In turn, those women who were able to withhold some of their earnings from their husbands reshaped household gender relations through this empowering, if highly risky, activity.

For encampment residents, access to alcohol was critical, as it played a central role in socialization and, as a libation, was instrumental in the (re)performance of certain rituals and ceremonies. In practice, there was little difficulty in securing alcohol. For example, because African women brewed "traditional" alcohol throughout Lunda, workers regularly purchased it in communities that they passed by during their daily commutes. As Caiombo Jombe explained, "We would often buy alcohol on the way home from work from women who would sell it along the road. So there could be drinking every night, but parties—drums, dancing, etc.—were only on Saturday nights, as there was no work on Sunday."[110] Diamang even allowed laborers in certain situations to purchase Portuguese wine in limited quantities from company stores, signifying that the enterprise was fully complicit in workers' consumption.

Workers and their wives also procured alcohol by illicitly trading their company-issued rations or clothes to neighboring villagers. According to João Paulo Sueno, "People from the encampment would exchange fuba or fish for traditional alcohol from the nearby villages."[111] Joaquim António Issuamo described a similar scenario regarding the procurement of alcohol, except his testimony reveals that individuals from local communities were more interested in company-issued clothing than rations. "We drank palm wine. . . . Wine was traded for in the [nearby] villages. . . . Food was not really traded for alcohol. Clothes for alcohol was a common trade."[112] Meanwhile, Mawassa Mwaninga explained how she and others in her encampment regularly engaged in this forbidden commerce, while lowering the risk level: "There was a lot of [illegal] trade between us and the nearby villagers. Money, clothes, food, wine, etc. all changed hands. Typically, we would trade money in exchange for things. All of this commerce was hidden; if whites saw, we would end up in prison [although here she laughed as a means of conveying the unlikelihood of this occurrence]. Actually, the sentinela often acted as an intermediary and took a fee for doing so."[113] In practice, not all encampment residents had sentinelas who were either this sympathetic or entrepreneurial. Yet as long as workers maintained high levels of professionalism during working hours, company officials were largely indifferent toward this form of post-shift commerce.[114]

Over time, a small number of women in company encampments opted to brew alcohol themselves rather than trading away valued food or clothing for it. Brewing is an endeavor that women have historically dominated both within Angola and across the continent, and female brewers were ubiquitous on and around mines throughout Africa. However, Diamang officials never tolerated the manufacture of alcoholic beverages within company settlements. Fearful that female brewing would jeopardize both the patriarchy and multidirectional fraternity that prevailed within encampments, company officials sought to inhibit this activity. Consequently, most informants dismissed the notion that women brewed in Diamang's mine encampments, declaring that it was impossible to pursue undetected and that because purchasing alcohol or simply trading for it was much easier, these latter options were preferable anyway.

Diamang's suppression methods were myriad and varied. Company officials ordered sentinelas, as their eyes in the encampments, to police brewing, which informants indicated that they did faithfully. According to Joaquim António Issuamo, "Beer and wine were not made in the house—[but] often near rivers, quite hidden. It was the sentinela's job to find them, but it was obviously difficult to do so. Cipaios were also responsible for finding them."[115] Diamang officials were also aware that otherwise innocuous items could be employed in the fermentation process. For example, in 1968, a company report warned of "empty fuel drums used to store water for construction purposes, but which could also be used as stills for the production and fermentation of alcohol."[116] It was this acute level of company concern and suspicion that greatly limited the numbers of women willing to manufacture alcohol in mine encampments.

Because of the risks involved, women who brewed in mine encampments were able to generate significant income. The revenues benefited both these women and their families, though also often included some wrangling with their spouses. For example, as Joaquim António Issuamo, proclaimed: "Women were the ones who made the alcohol, and as a result of their sales they would make more money than their husbands. But they were still inferior, and within the household the husband would always take the profits!"[117] Yet Lina Machamba, who brewed in an encampment associated with Chibale-Chomupila mine at the end of the 1960s, indicated to me that she maintained a "secret stash" derived from the brewing revenues that she was able to keep from her husband.[118] The monies amassed by Machamba and other brewers increased their autonomy while at Diamang and, in turn, helped them reshape their spousal relations.

STAYING CONNECTED: CULTURAL REPRODUCTIONS

Encampment residents did not leave their cultural beliefs or the ceremonies and rituals associated with these convictions behind in their home villages. Like songs, cultural re-creations were intended to emphasize continuity, reinvigorate links to distant homelands, and alleviate the experience of operating in an alien environment. To avoid the potential destabilization within the camps that so-called "detribalization" was perceived to sow, and also to preclude the development of "a dangerous proletariat," company officials uniformly encouraged these "traditional habits and customs."[119]

Adjudicative processes represented one type of cultural reproduction/performance in which, over time, workers participated. The first example from the written record comes from Parkinson and is related to the dispute from the 1920s (outlined above) that his Belgian co-worker was unable to resolve. "Both parties were badly dissatisfied with his decision and insisted on calling in a 'leopard judge.' When this was agreed to, the parties left, returning shortly with a huge leopard skin . . . starting weird incantations. They would pull out and eat a few hairs from the skin, circle the skin at top speed, chanting all the while. After about twenty minutes, they suddenly rolled up the skin and departed."[120] This incident underscores encampment residents' acceptance of recognized cultural traditions to resolve disputes either following unsatisfactory company mediation or, at other times, when none was immediately available. Dinis dos Santos Muriandambo, who worked for Diamang from 1969 to 1975, explained that when encampment residents were left to fend for themselves, a group of the *mais velhos* (eldest) would gather to settle disputes:

> The *mais velhos* in the encampment would gather to resolve problems in the same fashion a *soba* would . . . and then a fine would be levied against a person who, for example, went after another's wife. . . . These groups were composed of ten workers from the encampment, who most likely would not be as old as the elders back home in their villages. . . . It was the same group all the time regardless of the particular problem. They would pass around cups of alcohol, and when the container was finished, then the problem should be resolved. . . . This is very traditional. It was definitely a privilege to be in this group of ten. . . . We were able to practice our culture, ceremonies, etc., without any problems in the encampment. Diamang did not mind, nor did it cause friction between different groups.[121]

Muriandambo's testimony suggests that just as workers celebrated their culture(s) through familiar songs and dances meant to entertain, encampment residents also invoked culture and tradition when dealing with serious matters.

Even the absence of key authority figures in these reproductions failed to deter residents who were determined to engage in them. As part of the company's strategy to marginalize and supersede indigenous authorities, Diamang refused to employ, or even allow, sobas and traditional healers onto its mines and, therefore, these two key members of the village hierarchy were conspicuously absent from African laborers' ceremonial reproductions.[122] Yet workers ably improvised. Muriandambo further explained that "the oldest [men] in the encampment would assume the roles that would have been played by the elders in traditional ceremonies in the village. This went smoothly with no problems because everyone within that culture—it was in him [here he placed his hand up to his heart in the form of a fist to show that this 'culture' was deep inside each person]—so it was never difficult or foreign just because the ones who might traditionally assume these roles were not there."[123]

EXHILARATING LISTENING: NATIONALISTS' RADIO BROADCASTS

Workers' practice of furtively tuning in to the various guerrilla movements' nightly radio broadcasts to stay apprised of the latest martial and political developments constituted one of their riskiest undertakings. Owing to Portuguese employees' and family members' heightened fear of revolutionary violence, repercussions were severe and could even be fatal. In fact, most informants asserted that listening to these broadcasts was too dangerous or simply impossible after the arrival of the PIDE (in 1963), and thus they avoided even owning radios for fear of being accused of listening to these transmissions. Others, however, spiritedly admitted that they had tuned in. For these employees, listening to the broadcasts each evening helped temper the challenges that the next day would inevitably bring. Marissa Moorman has also compellingly argued that these broadcasts connected colonial subjects in Angola with the guerrillas fighting for independence beyond the border, helping to forge and shape a sense of an imagined Angolan nation.[124] Yet for all of the broadcasts' prosaic and inspirational value, there remained absolutely no unrest in Diamang's operational area during the conflict (1961–75), and very few Lunda residents ventured beyond the region to link up with the movements. Thus, both the action and *inaction* that African employees displayed collectively suggest that these individuals had reconciled their intolerance of Portuguese overrule with their tolerance of life as a Diamang employee.

The outbreak of the war for independence began at a particularly propitious moment in the history of communications. As late as the 1940s, generally only reasonably well-off Africans in the colony could afford radios, but by the 1960s and into the 1970s the proliferation of inexpensive transistor radios rendered them increasingly available.[125] Consequently, most informants indicated that radio ownership was viable, even if the company explicitly forbade workers to listen to the movements' broadcasts.[126]

For those employees bold enough to listen to the battlefield successes and other propaganda broadcast by the independence movements, discretion was vital. Joaquim Trinidade described this perilous process as follows: "The only subversive political activity post-1961 was clandestine, such as listening to the radio. You would only listen to it at home and were quiet afterwards and did not talk about it. 'Angola Combatante' was the subversive radio broadcast we would listen to . . . but, we would never talk about it, the PIDE, or the anticolonial liberation parties."[127] Other informants described a similar approach, noting that this level of caution was absolutely vital. As Silvestre Muachembe indicated, "If you were caught listening to 'Radio Brazzaville' or 'Angola Combatante,' you would get the radio smashed on your head, then you would be tied up and taken away and would disappear. You were a dead man."[128] Bartolomeu Lubano confirmed that this fate did, indeed, await those who were caught, but also indicated that he was so energized by the broadcasts that he ultimately left the company to join the MPLA (one of the Angolan independence movements).[129] "I worked for Diamang between 1965 and 1973, and then I stopped because of the war for independence. . . . I listened to the radio and was drawn in by the broadcasts. But, this was rare, as the PIDE would kill anyone caught listening, so people were scared. . . . My father had a radio and. . . . encouraged me to head in the direction of the movements."[130] For all his courage, Lubano's case is rather anomalous, as very few individuals who joined the movements emanated from Lunda. Instead, most workers who did engage in this nocturnal auditory activity on the mines did so primarily to help propel them through yet another long evening and subsequent workday.

☙

After the whistle blew each day, African employees returned to their respective living quarters. Once home, workers and/or family members prepared food, retrieved water, and socialized with friends and family members, among other activities. Although disputes occasionally disrupted mine encampments, the web of social relations in which individuals were entwined were generally cooperative in nature. Singing, dancing, and re-creating familiar ceremonies in the encampment, which each fed a broader process of cultural reproduction, promoted camaraderie and helped foster a degree of social harmony and stability that benefited both the residents and the company.

From the 1930s onward, conditions in mine encampments steadily improved. Diamang began upgrading housing, launched Spamoi, and achieved food security during this pivotal decade, though some African workers and family members continued to endure long commutes each day to and from mine encampments and were subject to various degrees of regulation once there. Yet, in general, conditions within Diamang's encampments continued to be far superior to those found elsewhere, both within the colony and across the region, and go a long way toward explaining the quietude that marked these settlements. Ultimately, Diamang's paternalistic policies and workers' social professionalism were responsible for generating and maintaining this durable placidity—even following the outbreak of the Angolan struggle for independence.

African laborers who successfully endured the working and living conditions at Diamang and, thus, reached the end of their written or oral contracts, were required to determine whether to return to (or in the case of most voluntários, remain in) their home villages or to stay on at Diamang. In the next chapter, I examine this decisive moment and the experiences that ensued for both the small number of workers who chose to stay with Diamang and the multitudes of employees who elected to re-trace their voyages to the mines in anticipation of long-awaited homecomings.

7 ⤙ To Stay or to Leave
The End of the Labor Contract, 1921–75

At the end of my two-year contract, I went back through Dundo. We spent a month there waiting for trucks to come pick us up, but we were not made to work. . . . After I reached my home posto, I went on foot. After I picked up my salary, I bought food for the walk home and clothes for my family. I also brought back things for the soba, who was my father. . . . But, the money was not sufficient. I had to work in the fields and also catch fish. . . . I got married and was home for a year and then returned voluntarily to Diamang because I liked it.

—*Former Diamang employee Rodrigues,*
describing the years 1958 to 1961

Life was better on the mines. We received money, clothes, and lots of different kinds of food. We didn't suffer so much. People preferred it.

—*Former Diamang employee Mulevana Camachele, who*
accompanied her husband to the mines in 1960

As THE COMPLETION OF African workers' contracts approached, they decided whether to extend their tenure with Diamang or to sever their engagement and return home. The company never pressured employees to stay on, allowing them to make this decision independently or, for those with accompanying wives, in conjunction with their spouses. For contratados, the prospect of staying on as a voluntário, and thereby likely receiving easier work assignments, enticed some to remain. However, most of these laborers declined, even if many of them would eventually be forced to return to the mines. Or would return voluntarily, as Rodrigues did, "because he liked it"; or as Mulevana Camachele suggested, because "people preferred it"; or simply because of material and/or survival considerations. Similarly, although more voluntários elected to stay on than did contratados, most voluntary workers still returned home for at least some period before re-engaging with Diamang. Over time, improved conditions and laborers' familiarity with occupational procedures and expectations translated to greater worker retention for Diamang, but the numbers of those employees willing to stay on following the end of their contracts remained frustratingly small for company officials. Although laborers exhibited a professional commitment to completing their daily tasks while on the mines, for most employees this approach constituted a strategic means to uneventfully disengage from the company, rather than to recommit to it.

For both voluntários and contratados, the process of returning home loosely resembled the course they had traveled to the mines, only in reverse. On reaching their home postos, contratados also collected a large portion of their salaries, which the company sent ahead as arrears to chefes do posto. Unfortunately for these laborers, in Diamang's early years, chefes often poached a portion of these monies, dispossessing workers of the very thing for which they had toiled so hard. Only as Diamang's power ascended vis-à-vis the state was the company able to effectively end this practice. From the posto, contracted workers and family members began the last stretch of the trip back to their home villages and the long-anticipated homecomings they typically received.

Workers remained in their home villages for varying amounts of time. Some returned quickly to the mines, forcibly or otherwise "recycled," while others' experience with Diamang was limited to a single contract. Over time, the tightening of the regional recruitment web forced an increasing number of workers to serve multiple stints with Diamang; improved conditions on the mines concurrently served to swell the numbers of returning voluntários. For company officials, experienced workers were attractive because they required less occupational training and time to (re)familiarize themselves with company policies and procedures. Consequently, Diamang introduced an array of incentives over time to entice existing employees to remain with the company, though most laborers opted instead for "repatriation," even if only temporarily.

In this chapter, I trace this "reverse journey" and the ways that workers and family members experienced it over time, from the initial decision to "stay or go" to their readjustments on returning home. Repatriation was less demanding than its mirror image—initial recruitment—and workers and family members readily extended their professionalism into the Lunda countryside, complying with policies and taking very few risks during the return process. Only strategic complaints issued to company officials about embezzling chefes do posto or, conversely, to these local colonial administrators about conditions at Diamang, interrupted workers' sheer determination to return home.

A (TEMPORARILY) UNAPPEALING OPTION: STAYING ON AT DIAMANG

Remaining with the company after the completion of a contract was one way that contratados could become voluntários and thereby possibly receive lighter postings, which also featured greater possibilities for advancement. As soon as Diamang began employing contratados, officials had hoped to achieve labor stabilization by expanding the ranks of these types of voluntários. Most contratados, however, disregarded company inducements. Similarly, even though voluntário workers could also improve their standing by staying on, company incentives were not compelling enough to entice most to immediately recommit. Over time, the numbers of workers who elected to stay increased, though they remained a fraction of the overall workforce. Mandatory returns to home villages before recommitting, obstructive local colonial officials, and occasionally hollow company assurances also impaired Diamang's retention strategies.

CONTRATADOS

Diamang officials never failed to cast instances in which even a handful of contratados chose to extend their tenure with the company as promising portents of future trends. A 1925 letter from Diamang's director in Dundo to its Technical Bureau in Brussels expressed this optimism: "We have received a few . . . families from Moxico during the past few months and hope to obtain more before the end of the year. We have had a number of them . . . complete their contracts and over 50 percent of these families have remained voluntarily in the mining area, and surely a good part of the contratados returning to Moxico will eventually return [to the company] as voluntários."[1] Similar to mining companies on the African Copperbelt, Diamang attempted to both attract and retain families as part of its overarching labor stabilization objectives, yet despite its efforts, it could not match the retention success that the Copperbelt companies enjoyed.[2]

As time passed, Diamang introduced a range of incentives intended to convince both single and married contracted laborers to recommit to the enterprise. By 1941, for example, the package on offer included company payment of the employee's annual taxes following the completion of eight months of service; admission to the second class/category of voluntários and, consequently, a right to regular salary increases; special encampments for ex-contratados in which they could cultivate crops and raise livestock; and a blanket and shirt.[3] Diamang even permitted some workers to remain in the immediate area without renewing their contracts and continued to cover their taxes for one to two years in the hopes that they would eventually re-engage with the enterprise in one form or another.[4] The company also often offered more agreeable work for those willing to stay on. Joaquim António Issuamo explained that "if contratados decided to remain with the company, they would not work on the mines again but received a different type of job, for example, in the hospital [as an auxiliary] or as a driver."[5] For Cafololo Gustavo, a former contratado, the tender was sufficiently alluring. "I signed on for another year after finishing my two-year contract [in 1961]. Next, I went to work as a domestic servant at the mine supervisor's house. . . . After 1962, I left the company and survived by working in the fields."[6] Gustavo's case notwithstanding, company figures reveal that despite the array of incentives on offer, Diamang failed to achieve any sort of widespread contratado retention (table 7.1). These low numbers perturbed company officials, who regularly touted the enterprise's "model" working and living conditions.

TABLE 7.1

Number of contratados staying on as voluntários by year

	1940	1941	1942
Ex-contratados who stayed on	22	18	25
Number existent at the end of the year	145	155	137
Total number of contratados employed	3,996	4,162	4,028

Sources: Letter from Jorge Figueiredo de Barros to the Snr. Governador Geral de Angola (March 11, 1943), 1, MAUC, Folder 86 42°; Letter from A. Pinto Ferreira to Ernesto de Vilhena (February 11, 1943), 2, MAUC, Folder 86D 3°; *Elementos sôbre mão-de-obra indígena* (1943), MAUC, Folder 86 43°.

The procedural path for those contratados who elected to remain with Diamang as voluntários was dissuasively circuitous. Instead of being allowed to stay in place, contratados were forced to travel at least as far as regional administrative centers—if not all the way back to their home postos—to formally declare their intention. A case outlined in a 1937 company report underscores the hazards involved in insisting on this procedure: "Although at the end of 1937, 193 ex-contratados declared that they wanted to stay on [the company referred to these individuals as "*prometidos*," or "promiseds"], only sixteen of these ex-contratados arrived back home and later returned to the mines to settle there."[7] The company even provided local sobas with gifts to persuade them to absorb these prometidos into their communities, as at this point Diamang was still not offering housing to voluntários, though these efforts met with only limited success.[8]

If reluctant sobas acted as potential obstacles, local colonial administrators more commonly impeded Diamang's labor stabilization objectives through a variety of obstructive measures. A company report from 1938 indicated that from October 1936 to April 1938, although 152 contratados had finished their contracts and had elected to stay, there were allegedly over 300 others who had desired to stay on but did not because proper arrangements could not be made with colonial officials.[9] A 1950 letter from a Diamang official in Dundo to company director Ernesto de Vilhena illuminates another way that local administrators hampered this process. "Since 1948, only three contratado workers have decided to stay on as voluntários. One of them, Tchibumbo, had asked for his family (his wife, mother, sister, brother and sister-in-law) to come, but despite repeated promises from the administrator of Songo, the official has not arranged for this to happen. Now, Tchibumbo wants to go home . . . which would not be the case if his family was with him."[10] In both instances, Diamang officials contended that colonial administrators were actively discouraging relocation so as to avoid losing a portion of their coveted tax base.

In rare cases, Diamang maintained the ranks of its experienced labor force by obliging workers whose contracts were expiring to remain on. Senior company officials were acutely aware of the negative impact this practice could have on both recruitment and desertion and thus they ensured that mine supervisors employed this retention scheme only during acute labor crises. In 1947, for example, alleged labor shortages surrounding the opening of two new mines and the urgent repairs necessary for a dam prompted one such instance. In this case, J. Tavares Paulo, the company's director general in Lunda, alerted local officials that he was invoking a clause in the colonial labor code (CTI) that allowed enterprises to extend workers' contracts beyond their originally scheduled termination date. Paulo's decision forced 459 contratados to remain at Diamang for an extra month, though they also earned a 25 percent salary bonus as a result of this imposition.[11] In 1957, the colonial state raised the legally allowable extension period to nine months, but also stipulated that workers had to consent to this labor elongation.

By the 1960s, company incentives to convince workers to recommit still had elicited little positive reaction from the African labor force. In part, this failure was due to a lack of creativity. Joaquim António Issuamo, a former employee, indicated that as of 1960 the core features of the company's original retention program were still firmly in place: a mandatory return to one's homeland and a potentially improved occupational position upon returning—all characterized by the same correspondingly low level of appeal among contratados. "Contratados who wanted to sign on for a second stint . . . had to return home first and then declare that they wanted to return."[12] In 1963, the company was even offering "workers from the exterior" increases in salary of up to 30–35 percent to continue as manual laborers, as well as payment of their taxes for up to five years. Yet, as with other incentives, these measures failed to entice many workers to remain with Diamang any longer than their contracts stipulated.

VOLUNTÁRIOS

Compared with contratados, voluntários' less dramatic relocation (if any) away from their home communities, as well as their shorter contracts, prompted many to extend their engagement with Diamang. Company incentives directed at voluntários were more aggressive, more creative, often more lucrative, and, ultimately, more effective than those offered to contratados. Fernando Tximvula, a former voluntário, revealed that Diamang even acceded to a transfer request he made in the late 1960s to retain his services. "I had completed two consecutive contracts, and then asked to relocate back to western Lunda because it was near my home, to which Diamang agreed."[13] Yet even though the numbers of voluntários who were willing to extend their employment with the company increased over time, most of these voluntary laborers spent at least some time away from the mines following the conclusion of their oral or written commitments before re-engaging with the enterprise.

Diamang's efforts to retain voluntários began shortly after it began employing these laborers and grew in scope over time. In 1925, for example, the company arranged for small food and/or monetary rewards to be awarded to voluntários who had completed 300 consecutive days (excluding Sundays) of service which, at that time, was roughly two to three times the average.[14] By the 1940s, Diamang had also begun to publicly commend voluntários who had both demonstrated "good behavior" and provided extended service to the company. These *antigos*, or veteran employees, were workers who had been with the company for over two decades. In 1945, for example, Diamang awarded a worker named Tchinzamba (either sixty-three or sixty-five years old at this point) with a monthly pension of 70 angolars for his dedication, as he had been with the company "even before Dundo had been founded in 1919."[15] By 1950, Diamang had lowered the threshold for antigo status to ten years. In that year, there were thirteen employees who had thirty or more years of service, thirty-eight with twenty-five or more, forty-three with more than twenty, and 250 with more than ten. At the annual Grande Festa (outlined in the prior chapter), Diamang distributed medals of copper, silver, and gold, as well as commensurate monetary bonuses, to these antigos. Veteran employees such as Tchinzamba typically entered as either contratado or voluntário manual laborers and over time worked themselves into less demanding (voluntário-only) occupations, such as a mechanic, carpenter, capita, shepherd, or driver. In doing so, they also served as regional advertisements for Diamang, even if they remained a small minority among the legions of voluntários who worked for the company over the decades.

At times, Diamang undermined its own efforts to retain voluntários by failing to fulfill promises it had made concerning the incentives it was offering. For example, in 1948 a company official reported that a group of voluntários who had provided extended service to the company (ranging from ten to fifteen years) was frustrated that they were still working on prospecting teams and that, despite their long tenures, they barely made any more in salary than newcomers and therefore wanted to transfer to lighter jobs.[16]

Into the 1960s, the array of factors outlined above continued to hinder the enterprise's labor stabilization objectives, ensuring that many more voluntários left than stayed (table 7.2). For every Tchinzamba

TABLE 7.2

Voluntários' tenure with Diamang as of 1968

Percentage with 1–2 years at Diamang	15.91%
Percentage with more than two years	23.91%
Total percentage with one or more years	39.82%

Source: António Botelho, "Nota de informação no. 6/70" (January 31, 1970), 16, MAUC, Folder 11-C (1970).

that Diamang hoped to generate, these powerful factors deterred most African laborers from (immediately) recommitting.

A BLISSFULLY CHALLENGING TRIP:
RETURNING HOME

For most contratados and family members who chose to return home, even a long journey back home was preferable to extending their time with the company. Although these individuals were essentially retracing their steps to the mines, this "repatriation" trip was easier for several reasons, including the introduction of mechanized transport in the 1930s, few if any labor requirements in Dundo, and, mentally, the prospect of returning home to friends and family after traversing what was now relatively familiar territory. Workers also looked forward to collecting the remainder of their salaries on reaching their home postos, even though this process was often fraught with problems. Consequently, although returning home was a challenging undertaking for most workers, it was an endeavor that they typically eagerly embraced. Voluntários similarly anticipated the end of their contracts, though they marked this event by unceremoniously walking home, just as they did virtually every other day.

BACK THROUGH THE "GARDEN CITY"

The first stop for most contratados returning home was Diamang's headquarters in Dundo. While there, workers and family members again stayed in the Transit Center where they mingled with incoming recruits for up to a month. Workers also underwent a simple medical exam that registered any weight fluctuations over their contractual period. In turn, Diamang used the largely favorable data to trumpet its treatment of the African labor force. Departing workers rarely performed tasks while in Dundo; instead, they simply waited to collect rations and then set out on foot or via truck transport.

The medical exam administered by company health services personnel was meant to screen for the usual battery of afflictions with which Diamang was concerned but also to measure any changes in weight experienced by workers over the course of their contracts. The company proffered the data generated by these measurements as evidence to rebut international claims that African laborers in Portugal's colonies were undernourished and/or overworked (table 7.3). If Diamang's figures are, in fact, accurate, they confirm that most workers were consuming a sufficient number of calories to more or less sustain their overall body weight. In general, the decreasing number of workers who lost weight over time reflects the improved food security

TABLE 7.3

Change in contratados' body weight during the contract period

	Gain of 1 kilogram or more	No gain	Loss of weight
1947	61.26%	18.27%	20.47%
1948	58.27	15.48	20.47
1949	46.16	17.76	36.08
1950	59.47	16.08	24.45
1951	56.70	15.65	27.65
1952	70.18	11.96	17.23
1953	37.75	16.18	46.07
1954	53.84	16.29	29.87
1955	54.99	17.10	27.91
1956	59.69	17.31	23.00
1957	72.23	13.12	14.65
1958	70.57	13.71	15.72
1959	64.55	14.78	20.67
1960	71.97	12.55	15.48
1961	63.53	14.67	19.80
1962	67.97	14.41	17.62
1963	69.85	13.08	17.07
Average	61.12%	15.20%	23.19%

Sources: José Picôto, *Assistência médico-cirúrgica na Lunda pelo Serviço de Saúde da Diamang: Elementos estatísticos de cinco anos de actividade, separata dos anais do Instituto de Medicina Tropical*, 10, no. 4, September 1953 (Porto: Imprensa Portuguesa, 1953): 2687; José Picôto, *Relatório médico annual* (February 27, 1954), 2, MAUC, Folder 86B, 6 6°; José Picôto, *Uma acção médico-sanitária em África altamente dignificante para Portugal: Assistência médico-sanitária na Lunda pela Companhia de Diamantes de Angola* (Porto: Costa Carregal, 1963), 51; José Henrique Santos David, "Regime alimentar dos trabalhadores efectivos da Diamang" (February 14, 1964), 4, MAUC, Folder 86, 2a 2°.

that Diamang achieved. The prevalence of weight gain may also be attributable to the increased number of calories that workers were consuming as compared to their pre-Diamang diets, as Noronha Feyo, a mine inspector, suggested in 1946. "The Chokwe, in general undernourished, reacts magnificently to the regime of nourishment that he encounters on the mines and in many cases he gains weight."[17] Workers' own efforts to supplement their food supplies also undoubtedly bolstered these figures.

While in Dundo, workers received rations for the return trip. Over time, the company gradually increased these allotments to make them consistent with the supplies distributed to recruits traveling to the mines. Diamang also periodically included "gifts" in these allocations, none of which were legally mandated. Beginning in 1940, for example, the company started distributing a blanket, a pano and a shirt to departing laborers. By 1945, accompanying wives were receiving a pano at both the point of recruitment and on repatriation.

Once outgoing workers had been processed, the trip home could begin in earnest. For approximately the first two decades of Diamang's operations, workers and family members completed this journey on foot. In the 1930s, however, the company strategically introduced mechanized transport for these repatriation journeys, roughly a decade earlier than it did for trips *to* the mines. Repatriating healthy workers by truck was meant to demonstrate Diamang's concern for its African labor force and thereby promote its regional image. Although potential recruits most likely saw through this calculated beneficence, returning workers and family members nonetheless benefited.

GETTING (PARTIALLY) PAID: AT THE POSTO

When workers arrived at their local postos, they anticipated collecting the portion of their salary, ranging from 50 to 100 percent, which colonial policy dictated that Diamang withhold.[18] The company opposed this legislation, fearful that local colonial administrators' (chefes do posto) practice of embezzling workers' salary arrears would prompt potential recruits to flee if they felt that no money would be left over after their tax obligations were deducted.[19] For their part, these colonial officials endorsed the withholding system so as to ensure that employees returned to their home areas and tendentiously contended that workers were "too irresponsible to retain enough of these large sums of money while on the mines in order to, later, pay their taxes."[20] Meanwhile, regardless of how senior colonial officials may have regarded local chefes' embezzlement schemes, the former did care about safeguarding the revenue stream that local taxes generated. Consequently, the colonial state stubbornly retained the withholding system throughout Diamang's operational era. Only by the 1940s, coinciding with Diamang's mounting profits and attendant ascension in power, do episodes of embezzlement disappear from the written record, and oral testimony confirms that by the 1950s incidents of this nature were extremely rare.

For returning workers, the arrears system was unfavorable at best. Although the officials in Lisbon who established colonial policy did not trust African workers to pay their taxes on returning to their respective localities, their skepticism should have instead been directed toward chefes do posto, many of whom used this process for self-enrichment. Common ploys were to exaggerate the annual taxes that Africans were obligated to pay, to fabricate debts that workers' relatives owed that returning laborers were suddenly responsible for covering, or to impose dubious late fees.

Prior to the company's expanding and increasing its regional power, unscrupulous chefes simply withheld monies at will and even threatened conscription into the colonial army for any returning contratados who protested. In 1921 at Villa Ariaga posto, for example, contratados returning from short, three-month contracts were due to receive 12$74 escudos apiece, but the chefe issued only 5$50 to each of them. When Diamang official M. K. Shaler intervened to try to rectify the situation, the contratados explained to him that if they objected, the state would have them arrested and sent away as soldiers. Shaler reported that the contratados added that "this was a common experience and they were powerless to protest and, therefore, did nothing as they preferred their liberty, such as it was, to life as a soldier."[21] A further incident, from 1930, reveals that this practice persisted, even if the threat of conscription was absent. In this case, at Lungué-Bungo posto, the chefe informed a repatriated group that they had worked, on average, only 35 "useful" (remunerative) days while at the mines during their six-month contract and that they would be paid accordingly.[22] Only after Diamang officials interceded, reinforming the chefe that this group had, in fact, worked 160 days during this period, was this situation redressed. A company report from 1938 illuminates another popular scheme employed by local administrators in which they required returning laborers to cover taxes allegedly owed by relatives or to pay concocted late fees associated with their own tax requirements. Either way, workers left the posto with little or nothing to show for their time on the mines.[23]

The overall system of withholding and then distributing salaries to local chefes do posto was almost as unattractive to Diamang as it was to its African employees. Although the company was hardly paying inflated wages to its workforce, Diamang certainly wanted what monies it did allocate for salaries to reach the appropriate recipients, rather than end up in colonial administrators' pockets. Over time, the company used its increasing power and influence to curb chefes' embezzlement schemes. Eventually, experiences such as the one conveyed in testimony provided by Luciane Kahanga related to his tenure at Diamang in the early 1960s became the norm: "When I arrived back here at the posto [Cafunfo] from Dundo, I received the rest of my salary that I was owed. There were no problems getting it. I paid my taxes and returned home with my friends."[24] As Diamang moved its complaints further up the colonial administrative hierarchy, the diamond enterprise gradually forced the compliance of these local officials. Naturally, the practice did not disappear entirely, but the paucity of incidents after the 1930s suggests that Diamang had effectively stifled it, and, thus, this iniquitous behavior would no longer undermine its paternalistic project.[25]

FULL CIRCLE: ARRIVING HOME

After workers had collected their after-tax wages at postos, they traveled over familiar terrain to reach their respective home villages. Arrivals were largely happy occasions for contratados and family members, many of whom had been gone for extended periods. Villages often commemorated these homecomings with feasts, while workers and family members reunited with and typically distributed gifts to friends and relatives and, occasionally, to sobas as well. Although one colonial official described the laborer's return as the moment "the indigenous contratado . . . comes home to find his lands demolished, his livestock missing, and many times his family life disorganized and undone," informants recalled that arrivals were instead generally joyous occasions.[26]

On returning home, some informants reported that they offered their sobas gifts, declaring that this gesture was very important. This sentiment underscores the fact that although these headmen had long since forfeited any meaningful authority, some of them were able to maintain a type of symbolic power that many young men continued to at least acknowledge. As Mateus Nanto recalled, "I returned home after my first contract in 1960 with a hat, jacket, shirt, and a pano that the company had given me, which all went to the soba."[27] Similarly, Cafololo Muamuiombo, who worked at Diamang intermittently during the 1960s and 1970s, explained that "we bought homemade clothes from neighboring villagers . . . and then we would bring these back to our families. The clothes we would buy were to wear, but also were things to bring back to the soba."[28]

In reciprocation for this bestowment of gifts, some sobas marked the occasion with a celebratory feast, which in Luciane Kahanga's case, who returned home from Diamang for the first time in the early 1960s, lasted for two days. Muhetxo Sapelende, recalling his experiences from the 1940s, added that "at times, and depending on their temperament, some sobas would thank the families who had sent their sons to the company. Upon the return of the worker, the family might receive a party in their honor, or a goat. But it was never given directly to an individual worker, just to the family."[29] Conversely, many other informants indicated that they did not bring anything back for their sobas, while residents of regional villages that were consistently visited by African police (*cipaios*) seeking recruits seemed somewhat inured to the arrival of returning laborers. As Domingos Cazeweque explained, "I did not bring gifts for my soba, nor did the soba throw us a party when my father and I returned from our contracts. This occurrence was so common that it hardly warranted a celebration."[30]

BACK TO WORK: CONTINUITIES AND CHANGES AFTER THE HOMECOMING

Because most workers' net, that is, after-tax, earnings were not sufficient to live on for any extended period, they were quickly forced to resume productive work to survive. As relatives typically tended to workers' (and wives') possessions and fields during their absence, most returnees simply resumed living as they had before leaving for Diamang. For most former mine workers, this meant eking out a living farming, fishing, and/or hunting, retaining enough to subsist and selling any surplus to either Diamang or the state. When comparing their pre- and post-Diamang lives, informants differed on how transformational their experiences at the company had been. Some noted few disparities, whereas others spoke of profound changes. In practice, many workers had multiple opportunities to make these assessments, as cipaios regularly rounded up already experienced recruits. Diamang coveted these "recycled" employees and thus the state recruiters regularly disregarded supposedly mandatory interval periods between labor contracts. Consequently, some returning workers had only weeks—or even days—to enjoy their homecoming before returning to the mines.

Regardless of how many times they served, many informants emphasized the transformational experience they had had at Diamang. Luciane Kahanga indicated, "After I returned, things were a bit different. I had more prestige in the village and was looked at a bit differently."[31] Others, such as Rodrigues (whose employment history was outlined in the epigraph, at the beginning of the chapter), indicated that despite the little money he had netted, their earnings had helped enable him and his wife to get married. And in one extreme case, António Muiege explained that following his return "I was chosen to be a soba, because [all] the elders in my village had died. I had the family link, of course, but I was also selected because I had knowledge of the Portuguese and other types of understandings because of my experience at Diamang. As soba, I was called to the posto by cipaios and the administrator would demand workers from me. I would then carry out the orders that I had received . . . in my village."[32]

Although some contratados emphasized the significant differences in their lives following their tenure with Diamang, others downplayed the impact. For Filipe Saucauenhe, for example, time spent on the mines effected few changes in his life. Regarding the return to his home village in 1967, he observed that "the suffering was the same. Nothing changed after I returned from my contract."[33] Other informants also emphasized these types of continuities. As Mateus Nanto proclaimed, "It [the Diamang experience] could not penetrate me. It could not and did not change my life."[34] In general, time spent at Diamang appears to have affected employees differently, and thus personal transformation should neither be presumed nor overemphasized.

For some workers, homecomings signaled the end of their employment history with Diamang, though for many others their returns concluded only one in a series of contracts into which they would enter, either voluntarily or forcibly. Unfortunately, company statistics are both spotty and speculative on this subject. Yet, they do offer a broad brushstroke of what this phenomenon may have looked like at certain junctures and generally indicate an increase in the numbers of workers on the mines who had previously served (tables 7.4 and 7.5).

TABLE 7.4

Percentage of workers who had completed a prior contract

Year	Percentage
1936	2.72%
1937	13.51
1938 (1st semester)	27.52
1940	20.30
1945	27.14
1947	30.53
1948	32.31
1949	40.78
1950	50.64
1951	49.54
1952	48.36
1962	43.00
1963	45.50

Sources: Diamang, *Contratados do Distrito da Lunda repatriados com contracto completo* (1938), MAUC, Folder 86 37°; *Notas sobre mão d'obra indígena* (August 1941), 5, MAUC, Folder 86 40°; MAUC, Folder 86B,6 5° (1953); Diamang, *Resumo dos trabalhos realizados em 1963. Relatório* no. 24 (nova série) 43 (1964), TT, Folder AOS/CO/UL-8A3, Pt. 1: 1940–64.

TABLE 7.5

Number of contracts that recruits from 1945 had served previously

Number of contracts	Number of recruits with that number of contracts
1	673
2	370
3	133
4	48
5	8
6	1
7	3
8	1
9	0
10	1
Total	1,238 (27.14%)

Source: Diamang, *Relatório da viagem do Dr. J. Simões Neves* (1945), 18, TT, Folder AOS/CO/UL-8A1.

For example, a 1938 Diamang report indicated that "many" contratados had by this time "already been contracted four times, and many others two and three times," and a 1953 company report determined that in 1951 "in order to maintain an average of 5,500 [contracted] workers on the mines, it had been necessary to procure 3,300 recruits during the year, of which *about* 50 percent had already been previously contracted, and to repatriate an equal number of them."[35] A report from 1961 is even more revealing concerning the imprecision of the data, but nonetheless captures the increase in the number of previously experienced workers in the company's employ: "We do not keep exact figures on those who claim that this is not their first contract, but would guess that . . . it is somewhere around 65 percent. This is all based on what the Africans tell us."[36] The inexactitude of the data notwithstanding, the figures suggest that "recycled" workers were increasingly prevalent on Diamang's mines especially as after the 1930s the company failed to broaden the areas from which it recruited in any significant way while continuing to require ever-expanding numbers of laborers. Consequently, as Diamang's operations matured (and conditions on its mines improved), workers were more apt to have served previously.

For workers who served multiple contracts, "resting periods" between contracts varied dramatically. Neither company nor colonial officials were overly concerned with how little time repatriated employees spent away from the mines. In response, some sobas implemented a type of rotational system to maximize rest periods for workers who had most recently served, while readily relinquishing subjects who had either never served previously or who were sufficiently rested. Costa Chicungo, who lived near Maludi mine, explained: "Actually, the process of who was selected was very fair. It basically worked on a rotational system. If you had already or recently gone as a contratado, then it was someone else's turn to go."[37]

Over time, as the company's labor demands grew, cipaios' visits grew increasingly frequent and intervals between contracts became harder, or even impossible, for sobas to safeguard. Moreover, colonial legislation introduced in 1952 shortened the long-standing official resting period from twelve to six months, though in practice the colonial state's recruiters widely ignored both of these intervals.[38] In 1962, for example, ILO investigators examining alleged recruitment and labor abuses in Portugal's African colonies interviewed a Diamang employee who admitted that the state often recontracted workers after as little as two months if there were shortages.[39] More egregious, though certainly rare, was Muhetxo Sapelende's experience in the early 1940s: "I returned to my village after my first contract, and then after only two days I was selected again!"[40]

Even after taxing stints on the mines and long journeys back to home villages, the unpredictable nature of the recruitment system afforded former employees little peace of mind. Many of those I interviewed had, after varying lengths of time away from the mines, eventually fulfilled multiple contracts. For example, João Muacasso was coercively contracted on multiple occasions, beginning in 1956. "I had four contracts. . . . I spent ten months on the mine the second time. I did this again and again for ten years. Over this period, the process remained the same—you still had to go to Dundo every time."[41] Other informants indicated that during the 1960s they returned willingly to Diamang after struggling through a hardscrabble existence at home. For example, Itela Joaquim indicated, "I . . . stayed home for two years, working in the fields, before coming back to work for the company. . . . I went back . . . because of the suffering in the village, the cost of living, and I needed clothes to wear, as did my wife and son. And, as I went on to have sixteen children, this was imperative!"[42] Others, such as Rodrigues, who returned voluntarily in 1969, as outlined in the epigraph, did so apparently because he "liked it."[43]

Some employees completed only one contract and, somewhat remarkably, neither volunteered nor were ever forced to return. Mulevana Camachele, describing her and her husband's sentiments after returning from their sole contract in 1961, proclaimed: "We never returned, never returned. I never wanted to return."[44] Conversely, Augusto Funete, who returned from Diamang in 1969, declared, "I never returned, but in my village I was not so secretive about my experience, and I even convinced others to go to the mines!"[45]

STRATEGIC RECOURSE: ISSUING COMPLAINTS AT THE POSTO

Repatriating workers and family members took very few risks during this process of disengaging from Diamang. By the end of their contracts, individuals had endured varying lengths of time on the mines and were now primarily focused on returning home, which the company was facilitating. Despite its appealing qualities, this portion of the journey was not completely devoid of challenges. In response, workers sought strategic solutions to the problems they faced, most notably by issuing complaints about disagreeable aspects of their employment to both company and colonial officials. Over time, this geographic extension of the "culture of complaints" that workers had previously cultivated on the mines met with similar success.

"IT WAS STEALING": COMPLAINTS ABOUT EMBEZZLEMENT

By far the most common predicament that returning workers faced was local colonial officials' practice of poaching a portion of their salaries. Although a handful of laborers angrily departed the scene of this injustice, thereby sacrificing any chance to recover their monies, most sought assistance from Diamang representatives in the field. Issuing complaints to these company officials constituted an effective strategy that carried only minimal risk, as these agents were eager to involve themselves in disputes of this nature. In particular, Diamang representatives were pragmatically concerned with both future recruitment and ensuring that the money paid out in wages reached its intended recipients. Consequently, by the end of the first two decades of Diamang's operations, workers' complaints, coupled with the enterprise's ascendant regional authority, helped eliminate local cases of embezzlement.

Returning laborers first began complaining to Diamang officials about corrupt chefes do posto in the early 1920s and continued to do so as long as chefes continued to embezzle salary arrears. In the incident at Villa Ariaga posto outlined above, workers brought the misconduct to the attention of company officials, just as they did in the later episode at Lungué-Bungo posto (also outlined above). In both cases, workers' complaints resulted in the recovery of purloined monies. Going forward, returning contratados had fewer misdeeds about which to complain—in great part due to their complaints. And by the end of the 1930s, cases of embezzlement had virtually disappeared.

AT A PRUDENT DISTANCE: COMPLAINTS ABOUT DIAMANG

Safely removed from the mines, workers also took the opportunity to complain about Diamang to local colonial administrators in the hopes that the latter would address at least some of the issues outlined in their grievances. Laborers' complaints typically detailed the challenging conditions on the mines, the repatriation process itself and the meager wages that they were in the process of collecting. In turn, chefes do posto used these complaints as ammunition during their ongoing regional power struggle with Diamang, that is, until the company's ascension effectively muted their protests.

Many of these complaints pertained to general conditions on the mines, especially during the early years of operations, before steadily fading over time as the company enhanced its range of services. Regardless, any grievances of this nature were worrisome to Diamang officials, who were preoccupied with the spread of bad publicity to either potential recruits or senior colonial administrators, or both. For example, in 1928, repatriated workers from Moxico issued a complaint to a chefe that ultimately reached both the colonial director of Native Affairs and the governor of Moxico. "We were made to start work before dawn . . . and were not relieved of service until six in the afternoon. The food was bad, consisting of, at times, only cowpeas and manioc . . . and the women often did not have time to transform it into fuba, and when fuba was distributed, it was often rotten."[46] These employees also revealed traces of whip marks and claimed that they had been beaten.

Another example comes from 1951, and was, thus, much more troubling to company officials, as improvements to living and working conditions had, by then, been under way for decades. Although seemingly isolated, in this case, fifty-seven repatriated workers from Songo complained to the chefe of Cambulo posto about indiscriminate violence on the mines, citing "African overseers who referred to us as vermin and told us that we ought to have our tails cut off," while also relaying to the chefe that two of their group had died while in service. The chefe indicated that these workers further explained to him that they "were unable to complain to a chefe do posto because their work site was very far from any posto . . . and that, after they rested for some time, they preferred to go to the Companhia de Assúcar (Sugar Company of Angola), where the treatment is much better than at Diamang."[47] For officials of the diamond enterprise, complaints emanating from this area were even more disconcerting because Songo had become a preferred recruiting district. Furthermore, Songo lay on the outer fringes of Diamang's exclusive recruitment zone and therefore residents could also be contracted by, or opt to work for, other enterprises, such as the sugar company. Beyond potentially losing the opportunity to recontract these fifty-seven experienced workers, the unfavorable commentary that they surely spread almost certainly cost Diamang other recruits, as well.

On returning to their local postos and retrieving the bulk of their salaries, workers also lodged complaints about the company's low wage levels (presumably only to those chefes do posto who had left their salary arrears unmolested). A colonial official's report from 1936 captures the nature of these complaints:

> The indígenas frequently complained that after they paid their taxes, nothing remained for them. It used to be that some of them, late in the payment of taxes, remained in debt to the state even after they had completed a contract. . . . There were times that at the moment at which the indígenas were repatriated . . . after they paid their outstanding taxes, they looked at the remainder, between 250 and 270 angolars, and told me: "So it is with this money that I am going to wait a year and a half until my field begins to give me manioc?"[48]

Even though Diamang's salary levels were comparable to those on offer from a series of enterprises that had operations in areas adjacent to Lunda, local colonial administrators bitter over the company's regional dominance made sure that these types of comments reached senior colonial officials in both Luanda and Lisbon.

Some workers also criticized the repatriation process itself. For example, in 1947, a group of thirty-five contratados returning to Minungo posto alerted the chefe that while they were transitioning through Dundo not only had they been made to work but that a capita named Mua-tchimica had extorted 60 angolars from each of them, threatening punishment for anyone who dared to resist. A concerned company recruiting agent, Carlos Sousa, who notified Diamang's director general, J. Tavares Paulo, of this incident, stated: "Through conversations that we had among them, we know that they intend to flee if they are obliged to return to Diamang to work as contratados, although many, regarding the company, indicated their satisfaction."[49] Although cases such as this one were rare, the incident underscores the willingness of any discontented contratados to complain openly about Diamang, once they had reached a safe distance.

⌒

Regardless of the length or nature of workers' contracts, on completion both contratados and voluntários had to decide whether to extend their service. For a variety of reasons, including, foremost, the challenging labor regime, most workers opted to return to their home communities, at least initially, and, if contratados, to reunite with family and friends after long periods of separation. Over time, an increasing number of both contratados and voluntários elected to stay with Diamang, but their numbers remained depressingly low for an enterprise that was allocating significant resources to upgrade conditions for its African labor force.

Although the trip home from Diamang virtually mirrored the journey to the mines, it differed in several important ways. The most formidable challenge for returning contracted workers was collecting their salaries when they arrived at their local postos. When improprieties surfaced during this process, laborers often complained to Diamang officials in the field to reclaim what was rightfully theirs. Spurred on by workers' complaints, after Diamang's mounting profits afforded it increased control over local colonial administrators, this practice virtually disappeared. "Repatriating" workers also took this opportunity to complain about Diamang to local colonial administrators who were, in turn, generally eager to collect and pass along these grievances as part of their local power struggle with the company. As outlined previously, state-company tension undermined cooperation between Diamang officials and local administrators in endeavors such as recruitment and the arrest of mine deserters, both of which required close collaboration between these two parties. Although returning workers engaged in comparatively less strategic behavior during repatriation owing to the minimally demanding nature of this segment of the broader Diamang experience, their extension of the "culture of complaints" into the Lunda countryside proved to be one of the most potent tactics that they employed throughout their entire engagement with the diamond enterprise.

Epilogue

We had fresh meat, clothes, cantinas, and food; we never lacked meat,
there was free medicine, and everything was affordable. It was a state
within the state. . . . The Diamang era was the last time whites and blacks
ate together.

—*António Sulessa, a former employee, reflecting*
on the "time of Diamang" in Lunda

We here suffered a lot, but the Portuguese taught us to work.

—*Bernardo Montaubuleno, a former employee*

WITH THE PASSAGE OF time, the gratitude-laced nostalgia for "the
time of Diamang"—exemplified in the excerpts that appear above—
is seemingly growing ever stronger. Irrespective of whether Sulessa's
remark about interracial dining is taken literally or metaphorically,
it's easy to comprehend why (t)his wistfulness perseveres. Following the departure of the Portuguese, the Angolan Civil War steadily
overwhelmed Lunda, plunging it into an enduring chaos that eventually produced the region's infamous "blood diamonds." Even after
the conclusion of the extended civil conflict in 2002 and the concomitant end of the "blood" or "conflict diamond" era in Angola,
fond memories of Diamang persist. In the following pages, I trace
the momentous events that transpired in Lunda following Angolan
independence and consider how these developments continue to
color reminiscences of the Diamang period. I also appraise the contemporary implications of the company's history and entertain the

question: What's currently at stake in thinking about Angola's diamond mining past?

In many ways, the history of diamond mining in the sovereign state of Angola began on April 25, 1974—the same date that Portugal's Estado Novo regime finally came to an end. On that eventful day, the durable dictatorship collapsed in a coup d'etat under the weight of the three African wars that the country was simultaneously waging in Angola, Mozambique, and Guiné (Guinea-Bissau). Motivated by an aspiration to halt the martial campaigns, the coup leaders, who were primarily midlevel military officers who had served in the African theater(s), made it clear that Portugal was no longer interested in maintaining its colonial empire.

The Alvor Accord consummated between Portugal and Angola's three factious liberation movements (the MPLA, FNLA, and UNITA) set November 11, 1975, as the date for Angolan independence and the withdrawal of any remaining Portuguese troops. However, the accord was inadequately designed, and internecine fighting continued between the movements, eventually transitioning into outright civil war.[1] This conflict proceeded more or less continuously for the next twenty-seven years and was fueled over time by varying degrees of Cuban, South African, Russian, and American involvement.

At the time of the coup, Angola possessed the largest number of people of European descent on the continent outside of South Africa. For most white residents of Angola, and the other Portuguese African colonies, the events in the metropole signaled the end of their colonial residency. The prospect of black majority rule in the territories prompted a mass exodus during which hundreds of thousands of so-called *retornados*, or returnees, migrated (back) to Portugal, while others fled to proximate, white-controlled territories such as Rhodesia (Zimbabwe) and South Africa. Diamang's white employees were no exception, and the company was quickly gutted of its core engineering and managerial staffs—positions it had never allowed African employees to occupy. Their flight, combined with the prospect of black rule, and the anticipated nationalization of Diamang's assets, portended the denouement of the diamond enterprise.

After April 1974, Diamang was closing entire mines with regularity owing to the departure of its technical personnel. As Daniel Catuesse, who worked for Diamang between January and November of 1975, explained, "At this time, many white bosses were leaving. Mines were closing left and right as whites departed, scared that they would go to prison if they stayed."[2] Many whites indicated that even being shot would be preferable to staying.[3] The attendant loss of production precluded the payment of salaries to the roughly 18,000 Africans who remained on the payroll, but who were, for all intents and purposes, inactive. Consequently, the labor force fell to nearly 6,000, and carat output plummeted from 2.4 million in 1974 to less than 350,000 in 1975–76.[4]

For Diamang's African laborers and their families, the coup in Portugal ushered in a period of initial exaltation, quickly followed by profound uncertainty and fear. Indeed, while it seemed as if the sun was finally about to set on colonial rule, the liberation movements had already begun to turn their weapons on one another. Further, Portuguese troops were still stationed in the colony, and Diamang continued to maintain its own security forces in Lunda to protect its ongoing, albeit dramatically contracted, operations. Bernardo Reis, the last Dundo-based Diamang director, provides an account of the resultant tension in the area during the unsettling days leading up to independence: "With the approach of independence, the populations of northeast Lunda burned some Portuguese flags. . . . Fearing that the same thing could eventually happen to other flags and could then progress into an uprising, we packed up any remaining flags, fastened a large stone to them and threw them to the bottom of the lagoon of Luachimo near the dam of the Dundo hydroelectric station. It was a difficult and especially significant moment in which the patriotic sentiment made a lasting impression and moved us all profoundly. Afterward, we left in silence with hardened hearts and tears running freely."[5]

Shortly thereafter, at sunset on November 10, 1975, the eve of Angolan independence, company officials in Dundo lowered the Portuguese flag for the last time.[6] Although Diamang's operations were by now crippled and its technical staff skeletal, the company survived as a corporate entity, largely intact until August 1977, at which point the Angolan state instituted an aggressive nationalization process.[7] Angola's first president, Agostinho Neto, began the declaration announcing the company's nationalization in acerbic fashion: "Throughout its fifty-six years of existence, Diamang . . . maintaining control of the production and sale of one of the principal riches of the people—diamonds—never once gave the Angolan people the opportunity to participate in the management of this wealth or to collect any justified profits from it."[8] In 1988, the corporate entity a shell of its former self, Diamang's few remaining assets were transferred to Endiama, Angola's parastatal diamond enterprise.

After having gradually encroached upon Lunda, the Angolan Civil War eventually engulfed the area, shattering the durable "Pax Diamanga." Going forward, the region was to become internationally notorious for supplying the "blood" or "conflict" diamonds that enabled the liberation-cum-rebel movement, UNITA, to fund its war efforts until its leader, Jonas Savimbi, was killed in a hail of bullets in February 2002. Yet well before Savimbi's death, Lunda had already devolved into a modern-day "wild west." With the civil war as a backdrop, the Angolan government, UNITA, and the hordes of immigrant diggers (*garimpeiros*) who had descended on the area were all aggressively vying for the riches embedded in Lunda's soils. During this tumultuous period in the newly independent nation's history, this resource-driven violence further destabilized the country both by helping to fund the civil conflict and by enticing migrants from across the continent to pour across Angola's borders in search of buried mineral wealth.

Although the end of the civil war in 2002 concluded one of these challenges, Lunda remains an unsettled space. The Angolan government continues to wrestle with the problem of immigrant miners, often conducting large-scale expulsion operations noted for their efficacy and brutality. Moreover, each time the Angolan government grants a new mining concession via Endiama, illegal operations in Lunda are forced to vacate operational areas to accommodate these new concessions; far removed from the capital, this process is rarely smooth or permanent. Given the financial stakes and the widespread prevalence of small arms, violence is virtually inevitable. In particular, the security forces that the multinational mining enterprises employ to safeguard their operations have raised the levels of violence and mistrust in the region, generating innumerable problems with the local population.[9] For most of these residents, the Diamang era never seemed so far away, or so good.

⌐

Decades after Diamang extracted its last diamond, it's small wonder that former employees remember the erstwhile "state within the state" so favorably. If the testimony by Sulessa and Montaubuleno that appeared at the beginning of the epilogue is indicative of local residents' sentiments, the Portuguese who lived and worked at Diamang remember it even more fondly—a truly blissful time and place. For example, Isabel Reis, who grew up in Dundo from 1960 until 1974, stated, "My generation . . . we very much had . . . a golden childhood. We were living in a 'golden cage.' The company provided our house and everything in it. We could ride horses, we had swimming pools, we had tennis, we even had cricket, which is a very British game! We had table tennis, we had games, we had *everything* there."[10] Echoing Reis, every Portuguese I interviewed spoke longingly of the "time of Diamang" and, apparently unwilling to disturb these pleasant, unsullied memories, insisted that they would never return to Lunda, or even Angola.

Beyond the considerable nostalgia associated with the colonial-era diamond enterprise, the Diamang experience constitutes more than just a distant memory periodically revisited by an aging set of Portuguese and African former employees. Indeed, this past has contemporary implications, having affected postcolonial Lunda and, by extension, Angola, in both immediate and enduring ways.

In the initial post-independence period, Diamang's decades-long demonstration that Lunda was capable of yielding substantial diamond wealth naturally attracted the attention of the rival Angolan nationalist movements. Consequently, the region became the site of some of the bloodiest battles of the ensuing, protracted civil conflict. Simply put, Diamang's considerable success fueled both the initial race for Lunda's bounty and the proceeding chaos that rendered the region a highly contested, dangerous place.

Following the conclusion of the Angolan Civil War in 2002, a new era of diamond mining was set to begin in the troubled country. Yet effectively controlling and maximizing profits from alluvial fields is something that has largely eluded post-independent African nations. In this sense, aspects of Angola's diamond mining past may well be contemporarily instructive, with lessons to offer both the modern state and mining capital as these entities seek to (stably) exploit the country's prodigious deposits.

The Diamang experience also continues to influence the way that Lunda residents think about nation building, resource exploitation, and human relations. When my African informants described the Diamang years much in the same way that some residents of the former Eastern bloc nations nostalgically recall the Communist era, they were certainly remembering the relative quietude and certainty of that time, but also how Diamang provided and steadily expanded remunerative opportunities for regional residents and acted in a reciprocal manner—demanding a great deal from its employees but also providing for them in many other ways. Recall the illustrative testimony that appeared in the Introduction from Rodrigues, who was urged to, and did, "work with força," and from Mawassa Mwaninga, who opted to stay on the job "all the way up until five days before [giving] birth." Both passages capture the constituent "demanding" and "providing" of this reciprocal relationship. My informants were also surely recalling the conciliatory ways that they and other indigenous employees on the mines cooperated and supported one another—in sharp contrast to the ensuing extended period of civil conflict in Angola in which mistrust, aggression, and divisiveness permeated domestic society.

In this respect, Diamang continues to serve as a type of model, imperfect as it may have been. And, justly or otherwise, the enterprise is against what local residents measure and compare the contemporary Angolan state's governance of, and resource (mis)management in, Lunda. As such, the enduring memories of Diamang—even if romanticized—continue to help undermine the Angolan government's attempts to build trust and national unity following the decades of devastating civil war. Locally, this durable sentimentality also helps generate, or perhaps simply exacerbate, discontent among Lunda residents who feel they have been neglected by the state and its coterie of commercial mining partners in the current grab for mineral wealth. In turn, many of them long, perhaps not so illogically, for a time when the cadences of Diamang's mine overseers and the revelry of Saturday evening encampment gatherings diffused across Lunda's rolling hills.

Notes

CHAPTER 1: AN INTRODUCTION
TO ANGOLA'S DIAMOND PAST

1. Investors included individuals and corporations from Portugal, France, Belgium, South Africa, and the United States, as well as the Portuguese state, which owned approximately 20 percent of the enterprise's shares. See chapter 2 for a detailed treatment of Diamang's corporate composition and organization.

2. The year 1975 serves as both a logical and analytical temporal parameter for this study, because although the newly sovereign Angolan state did not immediately nationalize the company, the civil conflict in the country largely paralyzed formal mining operations.

3. See, for example, Allen Isaacman, "Coercion, Paternalism and the Labour Process: The Mozambican Cotton Regime, 1938–1961," *Journal of Southern African Studies* 18, no. 3 (September 1992): 487–526. As Diamang's profits soared, it wielded increased influence over both local and metropolitan colonial officials, thereby further reinforcing its virtual autonomy.

4. Joaquim Trinidade, interview by the author, Lunda (Angola), November 15, 2004.

5. "Lunda" means "abandoned land" in Mbundu, the language in which the word was coined.

6. Rodrigues, interview by the author, Lunda (Angola), August 16, 2005. "Força," a word in Portuguese that is polysemous in English, in this case is meant to convey a sense of zeal that makes "effort," its direct English translation, inadequate.

7. Mawassa Mwaninga, interview by the author, Lunda (Angola), November 22, 2005.

8. This delineated area was known as the *zona unica de protecção*, or "unique protection zone."

9. This practice is exemplary of the type of "point to point" globalization that James Ferguson has considered, in which globalization "hops" from place to place, rather than "flowing" uniformly to reach all areas in its path. See James Ferguson, *Global Shadows: Africa in the Neoliberal World Order* (Durham, NC: Duke University Press, 2006).

10. For example, the hydroelectric plant still provides electricity to local residents, the airport still services the region (now, the province of Lunda

Norte), and Diamang's main hospital in Dundo continues to function in this capacity.

11. Patrick Harries, *Work, Culture, and Identity: Migrant Laborers in Mozambique and South Africa, c. 1860–1910* (Portsmouth, NH: Heinemann, 1994).

12. For example, Higginson's exploration of women's revenue-generating endeavors on the Copperbelt focuses on their subversive activities, including, most notoriously, smuggling banned items onto the mines. John Higginson, *A Working Class in the Making: Belgian Colonial Labor Policy, Private Enterprise, and the African Mineworker, 1907–1951* (Madison: University of Wisconsin Press, 1989). See also Charles Perrings, *Black Mineworkers in Central Africa: Industrial Strategies and the Evolution of an African Proletariat in the Copperbelt, 1911–41* (New York: Africana, 1979); Jane Parpart, *Labor and Capital on the African Copperbelt* (Philadelphia: Temple University Press, 1983); Jane Parpart, "The Household and the Mine Shaft: Gender and Class Struggles on the Zambian Copperbelt, 1926–64," *Journal of Southern African Studies* 13, no. 1 (1986): 36–56.

13. See, for example, Joseph B. Rukanshagiza, "African Armies: Understanding the Origin and Continuation of Their Non-Professionalism" (PhD diss., State University of New York, Albany, 1995); Boubacar N'Diaye, "Ivory Coast's Civilian Control Strategies, 1961–1998: A Critical Assessment," *Journal of Political and Military Sociology* 28, no. 2 (2000): 246–70.

14. *Shibalo* dictated that Africans had both a moral and a legal obligation to work and was codified through a series of colonial decrees beginning at the end of the nineteenth century.

15. Edward Alsworth Ross, *Report on Employment of Native Labor in Portuguese Africa* (New York: Abbott Press, 1925). Two other reports appeared prior to Diamang's operational era whose accusations are consistent with Ross's account, and similarly caused a great deal of international commotion. See Henry W. Nevinson, *A Modern Slavery* (1906; New York: Schocken Books, 1968); William A. Cadbury, *Labour in Portuguese West Africa* (London: George Routledge and Sons, 1910).

16. See, for example, Elizabeth Schmidt, *Peasants, Traders and Wives: Shona Women in the History of Zimbabwe, 1870–1939* (Portsmouth, NH: Heinemann, 1992); Charles van Onselen, *The Seed Is Mine: The Life of Kas Maine, a South African Sharecropper, 1894–1985* (New York: Hill and Wang, 1996); Jane L. Parpart, "'Wicked Women' and 'Respectable Ladies': Reconfiguring Gender on the Zambian Copperbelt, 1936–1964," in *Wicked Women and the Reconfiguration of Gender in Africa*, ed. Dorothy L. Hodgson and Sheryl A. McCurdy (Portsmouth, NH: Heinemann, 2001), 274–92.

17. See, for example, T. Dunbar Moodie, "Ethnic Violence on South African Gold Mines," *Journal of Southern African Studies* 18, no. 3 (1992): 584–613; T. Dunbar Moodie and Vivienne Ndatshe, *Going for Gold: Men, Mines, and Migration* (Berkeley: University of California Press, 1994).

18. These authors argue that miners regularly developed "brotherhoods" that were predicated on ethnicity, drinking habits, occupation, and/or tenure.

Robert J. Gordon, *Mines, Masters and Migrants: Life in a Namibian Compound* (Johannesburg: Ravan Press, 1977); T. Dunbar Moodie, "The Formal and Informal Social Structure of a South African Gold Mine," *Human Relations* 33, no. 8 (1980): 555–74; Dunbar Moodie, "Mine Culture and Miners' Identity on the South African Gold Mines," in *Town and Countryside in the Transvaal: Capitalist Penetration and Popular Response*, ed. Belinda Bozzoli (Johannesburg: Ravan Press, 1983).

19. José Coxi, interview by the author, Lunda (Angola), August 16, 2005.

20. It featured a one-hour time difference from the rest of the colony.

21. McNamara attributes violence on South Africa's gold mines not to "ethnic tensions or differences," but to "a lack of personal privacy of individuals crammed into these living quarters." Similarly, James attests that intergroup violence on these mines was due, in part, to the "poor conditions of the compounds." See J. K. McNamara, "Brothers and Work Mates: Home Friend Networks in the Social Life of Black Migrant Workers in a Gold Mine Hostel," in *Black Villagers in an Industrial Society*, ed. Philip Mayer (Cape Town: Oxford University Press, 1980), 320; Wilmot G. James, *Our Precious Metal: African Labour in South Africa's Gold Industry, 1970–1990* (Bloomington: Indiana University Press, 1992), 110.

22. See, for example, Jonathan Crush, Alan Jeeves, and David Yudelman, *South Africa's Labor Empire: A History of Black Migrancy to the Gold Mines* (Boulder, CO: Westview Press, 1991), 3. These authors contend that this wastefulness was due to the ever-expanding geographical pool from which African migrant laborers arrived on South Africa's mines.

23. A partial list of recent historical studies of Angola includes Isabel Henriques, *Território e identidade: A construção da Angola colonial, c. 1872–c. 1926* (Lisbon: Universidade de Lisboa, 2004); Emmanuel Kreike, *Re-creating Eden: Land Use, Environment, and Society in Southern Angola and Northern Namibia* (Portsmouth, NH: Heinemann, 2004); Marissa Moorman, *Intonations: A Social History of Music and Nation in Luanda, Angola, from 1945 to Recent Times* (Athens: Ohio University Press, 2008).

24. In fact, the various independence movements (MPLA, FNLA, and UNITA) had begun fighting each other prior to 1975. As such, the civil conflict predates Angolan independence.

25. For example, Nuno Porto has written on Diamang's museum, which was located in Dundo, hundreds of kilometers from the mines, but African laborers are virtually invisible in these otherwise valuable contributions. Nuno Porto, *Angola a preto e branco: Fotografia e ciência no Museu do Dundo 1940–1970* (Coimbra: Museu Antropológico da Universidade de Coimbra, 1999); Nuno Porto, "Manageable Past: Time and Native Culture at the Dundo Museum in Colonial Angola," *Cahiers d'Études Africaines* 39, no. 155 (1999): 767–87.

26. These records are stored, uncatalogued, in a small shed on the roof of the Anthropology building at the Universidade de Coimbra and are part of the collection of the Museum of Anthropology (MAUC).

27. See Todd Cleveland, "Appraising the Value of History: Fieldwork Strategies, Solutions and Lessons from Angola's Diamondiferous Lunda Region, 2004–6," in *Immigrant Academics and Cultural Challenges in a Global Environment*, ed. Femi Kolapo (Amherst, NY: Cambria Press, 2008), 239–62.

28. As of 2014, the Angolan regime has been in power for almost forty years and continues to steer Lunda's vast mineral wealth away from local residents, into both its own coffers and those of foreign mining operations.

29. Lunda was one of the regions in Africa that prompted the 2002 adoption of the Kimberley Process intended to stem the flow of these stones. The Kimberley Process outlines the provisions by which the trade in rough diamonds is to be regulated by countries, regional economic organizations and rough-diamond-trading entities so as to guard against so-called "blood" or "conflict" diamonds entering legitimate trade channels.

CHAPTER 2: A BOUNTIFUL PLACE

1. For further information regarding Portuguese military campaigns in Angola, see Réne Pélissier, *História das campanhas de Angola: Resistência e revoltas, 1845–1941*, 2 vols. (Lisbon: Editorial Estampa, 1986).

2. Joaquim António Issuamo, interview by the author, Lunda (Angola), November 19, 2004.

3. In 1938, Diamang's director, Ernesto de Vilhena, speculated to António Salazar, the Portuguese dictator, that "unlike other metals [gold, copper, etc.] in Africa that the Africans had some knowledge of and used, diamonds in Angola were unknown to Africans" (Letter from E. de Vilhena to A. Salazar [February 9, 1938], 61, Arquivo Torre do Tombo [hereafter TT], Folder AOS/CO/UL-8A2). A 1923 attack on a group of African employees transporting diamonds from Luaco mine to Dundo by a band of regional residents hostile to Diamang, during which the assailants stripped and whipped the couriers, but ultimately allowed them to proceed with the stones, suggests that de Vilhena may have been correct (Letter from G. H. Newport [December 31, 1923], 11, Museum of Anthropology, Universidade de Coimbra [hereafter MAUC], Folder 86 2°). Regardless, throughout the Diamang era, rumors persisted that local populations had sold or incorporated diamonds into various items during precolonial times. A partial list of rumors included that residents from western Lunda used to sell diamonds to Portuguese merchants; that regional huts were adorned with stones; and that sobas buried them near their huts.

4. For further information on sertanejos, see Castro Soromenho, *Sertanejos de Angola* (Lisbon: Agência Geral das Colónias, 1943); José Carlos de Oliveira, *O comerciante do mato: O comércio no interior de Angola e Congo* (Coimbra: Centro de Estudos Africanos, 2004).

5. For information on this trade and coeval regional events, see Joseph C. Miller, *Cokwe Expansion 1850–1900* (Madison: African Studies Program, University of Wisconsin, 1969); Jeremy Ball, "'A Time of Clothes': The Angolan Rubber Boom, 1886–1902," *Ufahamu* 28, no. 1 (2000): 25–42; Ana Paula Ribeiro Tavares, "História e memória: Estudo sobre as Sociedades Lunda e

Cokwe de Angola" (PhD diss., Universidade Nova de Lisboa, 2009). As testament to the importance of rubber in the area, the Chokwe word for rubber is "dundo," the name of the site of Diamang's headquarters in Lunda.

6. For an account of labor migrants from Lunda working in Northern Rhodesia, see Charles Perrings, "'Good Lawyers but Poor Workers': Recruited Angolan Labour in the Copper Mines of Katanga, 1917–1921," *Journal of African History* 18, no. 2 (1977): 237–59.

7. Ibid., 238. According to Ball, "The producers and traders in Angola made no efforts to conserve or re-plant sources of raw rubber. The rubber trade followed an historic pattern in Angola of extractive commodity export until the point of exhaustion." Ball, "Time of Clothes," 40.

8. Distrito da Lunda, Ano Economico de 1915–16 (August 31, 1916), 3, Arquivo Histórico Nacional de Angola (hereafter AHN), Caixa 3510 "Malange."

9. Todd Cleveland, "The Life of a Portuguese Colonialist: General José Norton de Matos (1867–1955)" (MA thesis, University of New Hampshire, 2000), 42.

10. Copia de nota N. 447 da Capitania-Mor de Camaxilo de 10 de Dezembro de 1917 to the Secretatrio do Distrito da Lunda (December 17, 1930), 1, AHN, Caixa 3348, Camaxilo.

11. PEMA was formed on September 4, 1912, and headquartered in Lisbon. See Bernardo Reis, ed., *Diamante*, XXII Encontro Diamang, Tomar, June 12, 2004 (Braga. 2004): 10.

12. For example, PEMA's first prospecting foray in Lunda resulted in the death of one of the two lead engineers following an ambush by a hostile Chokwe community. Lute J. Parkinson, *Memoirs of African Mining* (self-published, 1962), 38.

13. Portuguese officials were, however, well represented on Diamang's Administrative Council.

14. PEMA had already opened its first mine the previous year along the Chihumbe River. In addition to compensating PEMA for its concessionary rights, Diamang also reimbursed PEMA for this undertaking.

15. Portuguese escudos were used in Angola from 1914 to 1928 and again from 1958 to 1977. From 1928 to 1958, the Angolan angolar was used, which was exchanged at a ratio of 1:1 with the Portuguese escudo.

16. The Republican government even attempted to institute measures to curb forced labor or otherwise offer limited protections to African workers, yet commercial interests in Angola blatantly flouted these halfhearted labor regulations. Portugal's meager administrative resources prevented it from trying to enforce much, if any, of this legislation, anyway.

17. M. Anne Pitcher, "Sowing the Seeds of Failure: Early Portuguese Cotton Cultivation in Angola and Mozambique, 1820–1926," *Journal of Southern African Studies* 17, no. 1 (March 1991): 62.

18. Newitt explains that "by 1925, the debt of Angola alone was over £4,500,000, a paltry amount compared to Portugal's £80,000,000 debt, but annual debt charges were consuming half of the country's income, which had

scarcely risen . . . as a result of Norton de Matos' speculations." Malyn Newitt, *Portugal in Africa: The Last Hundred Years* (Harlow, Essex: Longman, 1981), 178. That same year, Norton de Matos fell shy of the £13,000,000 required for a loan repayment to English investors, and the Lisbon government was forced to intercede.

19. Infrastructure improvements included the completion of the Benguela railway in the early 1930s, which linked Angola's coast with the interior of the continent, and, later on, the construction of regional airports.

20. Many of these colonists arrived as part of rural settlement schemes, in which the state set aside land in the colony's fertile central highlands. Over time, though, most of these settlers migrated to Angola's cities, principally Luanda. The estimated numbers of Europeans in Angola were, by year: 1900, 9,177; 1920, 20,000; 1940, 44,083; 1950, 78,000; 1955, 110,000; 1960, 172,000. See Douglas L. Wheeler and René Pélissier, *Angola* (New York: Praeger, 1971), 138.

21. In 1946, revenues from coffee exports overtook those from diamond exports. And, in 1973, coffee was itself surpassed by the exports of the incipient petroleum industry. Carlos Rocha Dilwa, *Contribuição à história económica de Angola* (Luanda: I.N.A., 1978), 267.

22. In 1959, in response to the mounting international criticism of labor practices in its colonies, Portugal ratified the ILO's Abolition of Forced Labour Convention. Shortly afterward, the newly independent state of Ghana (formerly, the Gold Coast colony) filed a complaint via the ILO against Portugal, accusing it of violating the Convention. Interestingly, the Ghanaian state cited Basil Davidson—journalist, scholar, and harsh critic of the Portuguese colonial project—as one of its witnesses.

23. Edward Alsworth Ross, *Report on Employment of Native Labor in Portuguese Africa* (New York: Abbott Press, 1925); Henrique Galvão, *Exposição do deputado Henrique Galvão, à comissão de colónias da Assembleia Nacional, em Janeiro de 1947* (Lisbon: Arquivo Histórico-Parliamentar, Assembleia da República, 1947); International Labour Office, Report of the Ad Hoc Committee on Forced Labor, ILO, New Series of the ILO (Geneva, 1953), Memorandum of February 22, 1952; Cunha Leal, *Coisas do tempo presente, I: Coisas da Companhia de Diamantes de Angola* (Lisbon: Edição do Autor, 1957); Cunha Leal, *Coisas do tempo presente, II: Novas coisas da Companhia de Diamantes de Angola* (Lisbon: Edição do Autor, 1959); Cunha Leal, *Coisas do tempo presente: Peregrinações através do poder econômico* (Lisbon: Edição do Autor, 1960).

24. ILO Official Bulletin, # 45 (XLV) 1962, supplement II, no. 2, April 1962. *Report of the Commission Appointed under Article 26 of the Constitution of the ILO to Examine the Complaint Filed by the Government of Ghana concerning the Observance by the Government of Portugal of the Abolition of Forced Labor Convention, 1957 (no. 105)* (ILO: Geneva, 1962); International Labor Conference, 56th Session, Geneva 1971. Report III Part 4A. *Report of the Committee of Experts on the Application of Conventions and Recommendations* (ILO: Geneva, 1971).

25. Letter from Guilherme Moreira, o administrador-delegado de Diamang, to Contra-Almirante Vasco Lopes Alves (February 8, 1967), MAUC, Folder 84B,10 1°.

26. ILO Official Bulletin # 45, *Report of the Commission*, 211.

27. Letter from A. Brandão de Mello to Norton de Matos (April 24, 1921), 3–4, MAUC, Folder 84K3 1°.

28. This arrangement led company officials to occasionally refer to *contratados* as *trabalhadores contratados com intervenção da autoridade*, or "workers contracted with the intervention of the authorities," and *voluntários* as *trabalhadores contratados sem intervenção da autoridade*, or "workers contracted without the intervention of the authorities."

29. Letter from the Governor of Lunda to Diamang (October 19, 1929), MAUC, Folder 86 18°.

30. Letter from James R. Evans to Diamang (July 15, 1930), 1, MAUC, Folder 86 20°. According to Heywood, "Taxes paid by Africans provided 20% of the colony's budget and represented the single largest source of revenue. Most of this income came from the taxes the employers deducted directly from the wages of contract laborers, not from income earned by Africans in independent economic activity." Linda Heywood, "Slavery and Forced Labor in the Changing Political Economy of Central Angola, 1850–1949," in *The End of Slavery in Africa*, ed. Suzanne Miers and Richard Roberts (Madison: University of Wisconsin Press, 1988), 430.

31. Letter from Jorge Figueiredo de Barros to V. L. Alves (December 5, 1934), 1, MAUC, Folder 86 27°. This letter triggered a series of others, in which various company officials cited additional posto vacancies.

32. Letter from Ernesto de Vilhena to António Salazar (July 20, 1963), 5, TT, Folder 20A 7°. In a subsequent letter, de Vilhena again appealed to Salazar regarding the governor general, complaining that he "molests, offends and damages us, always finding a reason, though unfounded, to do this, with the pretext of establishing for us a new regime . . . one which aims . . . to demolish an institution that he still does not know or understand . . . . and apart from that, he has a lack of knowledge and overseas experience." Letter from Ernesto de Vilhena to António Salazar (June 11, 1964), 1, TT, Folder 20A 8°.

33. Joaquim Trinidade, interview by the author, Lunda (Angola), November 15, 2004.

34. Eugenio de Barros Soares Branco, *Relatório do governador do distrito da Lunda* (September 1916), 52, AHN, Caixa 3894, Lunda.

35. These items increased in value over time as a direct result of the state's procurement of laborers. Thus, as the state rounded up workers for the mines, it was also indirectly helping itself financially.

36. Upon granting the colony 100,000 shares, Diamang increased the overall number of shares in the company from 800,000 to 2,000,000, though it maintained that individual shares would retain their value (4.5 escudos). See Diamang, *Estatutos da Companhia de Diamantes de Angola* (Lisbon:

Neogravura, 1939), 25; Diamang, *Estatutos da Companhia de Diamantes de Angola* (Lisbon: Oficina Grafica, 1950), 43–44.

37. Diamang, *Estatutos da Companhia de Diamantes de Angola* (Lisbon: Oficina Grafica, 1950), 37; Basil Davidson, *The African Awakening* (Oxford: Alden Press, 1955), 213.

38. Douglas L. Wheeler, "José Norton de Matos (1867–1955)," in *African Proconsuls: European Governors in Africa*, ed. L. H. Gann and Peter Duignan (New York: Free Press, 1978), 453.

39. Loans and amounts conceded to the Angolan Government include: 1921–22: £400,000; 1923–35: £397,360; 1937–39: £250,000; 1947: 100,000,000 escudos; 1955: 100,000,000 escudos; 1961–62: 105,620,000 escudos; 1963: 150,000,000 escudos. For further details, see Diamang, *Breve noticia sobre a sua actividade em Angola* (Lisbon: Tip. Silvas, 1963), 12–15, 103.

40. Diamang, *Notícia succinta sobre a sua constituição, concessões obtidas e trabalhos realizados em Angola* (Lisbon, 1929), 49. Assisting Diamang in the generation of these lofty revenues was the high percentage of "gem-quality" stones emanating from Lunda—second only to Southwest Africa (Namibia). Conversely, lower grade stones are generally used for industrial purposes. In addition to the influence with the state that Diamang's contributions to state coffers generated, the enterprise also wielded considerable power in the colony due to the fact that it was the *sole* exporter of diamonds; in Angola, *Diamang* and the *diamond industry* were synonymous terms. Thus, although the value of Diamang's yearly output often lagged behind the total values of the colony's coffee and maize exports, other industries—including both coffee and maize—were composed of hundreds, if not thousands, of producers, whose individual power and influence was diluted and, thus, relatively minimal.

41. Portuguese commanders cited Calendende's poor leadership and tactical skills as the primary reasons for his eventual defeat. For more information concerning local Portuguese military endeavors, see Alberto de Almeida Teixeira, *Lunda: Sua organização e ocupação* (Lisbon: Agência Geral dos Colónias, 1948), 242.

42. Letter from Brandão de Mello to Vasco Lopes Alves (December 17, 1935), 2, MAUC, Folder 86 29°.

43. Letter from Ernesto de Vilhena to António Salazar, "Angola de Couto Rosado" (February 9, 1938), 171, TT, Folder AOS/CO/UL-8A1. These two examples come from 1922 and 1926, respectively.

44. Letter to Snr. Col. António J. Santa Clara from António Brandão de Mello (June 24, 1921), 2, MAUC, Folder 84K3 1°. Cunza was also known as Gunza and N'Gunza. Just two years earlier, Cunza had apparently conveyed to Diamang authorities that "he had no complications with the Company, and, in fact, wished for Diamang to settle in Lunda and to continue work, but that he was determined to oppose the Portuguese soldiers with all his force." Letter from R. B. Oliver, Managing Engineer, to the Société Internationale Forestière et Minière du Congo, Congo Belge (October 28, 1919), MAUC, Folder 84K3 1°.

45. Letter to Norton de Matos from A. Brandão de Mello (April 24, 1921), 4, MAUC, Folder 84K3 1°.

46. Letter to Lt. Col. Duarte Silva from G. H. Newport (January 15, 1921), 1, MAUC, Folder 84K3 1°.

47. Parkinson worked on diamond mines in the Belgian Congo and South Africa, in addition to Diamang. His *Memoirs* chronicles his time with Diamang in the 1920s and 1930s, and also a return trip to the area he made some decades later. Although often racist, his insights into daily life on and around Diamang's mines remain extremely valuable. Lute J. Parkinson, *Memoirs of African Mining* (self-published, 1962), 39.

48. A reasonably large, though isolated, uprising in October 1931 in the region caused concern in both company and colonial circles. According to a government report, "Troops sent . . . resulted in the deaths (or injuries) of some rebels. In the village of the soba Mona Quimbundo, leader of the revolt, who still has not been apprehended, there were gathered approximately 5,000 natives. Some were taken prisoner and a great quantity of gunpowder and fuses were confiscated." Letter from J. Salema Vaz to the Snr. Representante da Diamang, "Insubordinação em M. Quimbundo" (September 18, 1931), 1, MAUC, Folder 86 23°.

49. Letter from Ernesto de Vilhena to António Salazar, "Angola de Couto Rosado" (February 9, 1938), 172, TT, Folder AOS/CO/UL-8A1. Amazingly, de Vilhena guided Diamang from Lisbon from 1919 until shortly before his death in 1967. In 1938, Brandão de Mello boasted to the visiting Minister of the Colonies Vieira Machado that "in case it is necessary, the colony can count on us and our workers and personnel to mount here, with some 15 days of preparation, an organized military force of 10,000 blacks, along with 100 whites, ready to defend the integrity of these territories, with all the ardor and enthusiasm that you noted." *Boletim Geral das Colonias.* no. 162 (December 1938), 542.

50. Diamang, *Acta da 55.ª Reunião da assembleia geral (ordinária) dos accionistas da Companhia de Diamantes de Angola, Realizada no dia 30 de Maio de 1960* (Lisbon: Tipografia Silvas, 1960), 29.

51. By 1963, Diamang was bemoaning the costs associated with the construction of barracks for regular army troops in the area that it estimated at over 50,000,000 escudos (£625,000). Moreover, it claimed to have spent over 37,000,000 escudos by this time arming its own Corpo, which had grown to approximately 200 men. Diamang, *Acta de 58th reunião da assembleia geral (ordinária) dos accionistas da Companhia de Diamantes de Angola, realizada no dia 28 de Junho de 1963* (Lisbon: Silvas, 1963), 18.

52. In great part, this speculation was based on company and government informants' accounts of UPA's intentions, which apparently included a plan to overtake the company by first seizing the main airport in Dundo. Letter from António Baptista Potier to the Policia Internacional e de Defesa do Estado (January 16, 1963), TT, Folder PIDE/DGS Del A, P Inf, 11.27.D/1. UPA eventually became the durable, yet largely politically and martially ineffective, Frente Nacional de Libertação de Angola (FNLA).

53. Interview, Costa Chicungo, Lunda, May 13, 2005. Ball indicates that the PIDE worked with the Cassequel Company in Angola in a similar manner. Jeremy Ball, "'The Colossal Lie': The Sociedade Agrícola do Cassequel and Portuguese Colonial Labor Policy in Angola, 1899–1977" (PhD diss., UCLA, 2003), 237.

54. It was UNITA's ongoing usage of "blood diamond" profits to fund its campaign in the Angolan Civil War (1975–2002) that helped prompt the creation of the Kimberley process. There is ample evidence that both Diamang and the state were concerned that nationalist movements would try to fund the revolution using diamonds from Lunda, but no evidence exists that these groups ever succeeded in doing so. See, for example. Letter, Carlos Eduardo Machado (February 22, 1967), TT, Folder AOS/CO/UL-8A4 Cont. Although an examination of why FNLA chose to concentrate its attacks elsewhere on the Angolan-Congolese border is outside of the scope of this book, former Diamang security chief Leonardo Manuel Judas Chagas indicated that he was in regular contact with UPA officials and that an attack on Lunda was never even a remote possibility. Leonardo Manuel Judas Chagas, interview by the author, Lisbon, June 7, 2006.

55. Leonardo Manuel Judas Chagas, interview by the author, Lisbon, June 7, 2006.

56. José Silva, interview by the author, Lunda (Angola), August 12, 2005.

57. Parkinson, *Memoirs of African Mining*, 3–5. The fact that the massive Société Générale de Belgique conglomerate was an important investor in both firms further solidified the relationship.

58. By 1929, Diamang had already virtually matched Forminière's output, before permanently surpassing it in 1937. Approximate output in carats: 1917: Forminière (F)—100,000, Diamang (D)—4,000; 1919: F—215,000, D—49,000; 1920: F—213,000; D—94,000; 1929: F—325,000, D—312,000; 1930: F—339,000, D—330,000; 1937: F—568,000, D—626,000. Correspondência/SALAZAR, n.º 79, Anexo 1, TT, Cota: Arquivo Marcello Caetano (hereafter AMC), CX. 51; "Comparação da produção, 1912–1947" (1948), TT, AMC.

59. As in most Bantu languages, the prefix for "people" is *ba*-, and thus *Baluba* refers to the Luba peoples and is analogous to, for example, "Americans," while *Luba* is used as an adjective, e.g., Luba workers.

60. Manuel Figueira, *Relato sôbre mão d'obra indígena* (May 18, 1938), 2, MAUC, Folder 86 36°. For more information on these employees, see Todd Cleveland, "A Minority in the Middle: Ethnic Baluba, the Portuguese Colonial State, and the Companhia de Diamantes de Angola (Diamang)," in *Minorities and the State in Africa*, ed. Michael Ubanaso and Chima Korieh (Amherst, NY: Cambria Press, 2010), 195–216.

61. Diamang, *Notícia succinta sobre a sua constituição*, 23. The average number of African employees in the first two years of mining operations was 593 in 1917 and 830 in 1918.

62. V. Matos, "Pioneiros," in B. Reis, ed., *Diamante*, XXI Encontro Diamang, June 2003 (Braga, 2003): 5.

63. Included in this process was the symbolically important, initial assumption of the head technical post in Lunda by a Portuguese, the engineer Quirino da Fonseca, in 1934, ten years after he had arrived. *Algumas notas sôbre o pessoal branco das explorações* (1934), 1, TT, Folder AOS/CO/UL-8A. Decree 241 of February 20, 1923, obliged concessionary companies to have a Portuguese in their highest-ranking position in the colony, Portuguese in at least half the technical, or administrative and accounting, positions and in all medical and nursing posts. This legislation extended an earlier, largely ignored provision in the 1921 state-company agreement stipulating that 70 percent of the European (white) personnel at Diamang had to be Portuguese, along with the representative to the colonial government, the personnel employed in the repression of (illicit) diamond traffic, the recruiting agents, and the technical director of mining operations.

CHAPTER 3: THE RECRUITMENT
PROCESS, 1921–75

1. Conveção entre Sr. Governador da Lunda e O Rep. da Diamang (1919), Museum of Anthropology, Universidade de Coimbra (hereafter MAUC), Folder 84K3 1°.

2. *Boletim Oficial da Província de Angola*, series 1, no. 20 (May 14, 1921): 37.

3. Company officials regularly fixed Lunda's population density at two people per square kilometer, though censuses were inaccurate and this figure was bandied about well before the first census was ever taken.

4. Manuel Joaquim de Magalhaes, repartição distrital de administração politica e civil, provincia de Angola, C. de Fronteira do Chitato, Dundo (April 15, 1924), 3, Arquivo Histórico Nacional de Angola (hereafter AHN), Caixa 3932, Lunda.

5. Letter from João Augusto Bexigo, Director Geral na Lunda, to Diamang (Lisboa), "Notas de informação nos. 43 e 44/71 sobre rendimento da mão-de-obra" (July 21, 1971), 1–2, MAUC, Folder 84G,1 3.

6. Joaquim António Issuamo, interview by the author, Lunda (Angola) November 19, 2004.

7. For example, in 1941 the state transferred a group of Mucubais to Diamang consisting of 90 men, 93 women, and 70 children. Their POW status may have enabled the company to retain their services longer as they were allegedly "excellent employees." See J. Almeida Santos, "Trabalhos Gerais" (January 20, 1945), 1, MAUC, Folder 86 45°; Letter from A. Pinto Ferreira to Jorge Figueiredo de Barros "Prisoneiros Mucubais" (July 17, 1945), 1, MAUC, Folder 86 46°.

8. Diamang's mobile health teams traveled to villages throughout Lunda and, among other endeavors, actively attempted to stamp out leprosy and trypanosomiasis, or "sleeping sickness," and, to a lesser extent, tuberculosis and malaria. These teams would also distribute food and blankets as part of its pragmatic effort to build regional trust and goodwill toward the company.

9. ILO Official Bulletin, # 45 (XLV) 1962, supplement II, no. 2 (April 1962): 38.

10. Henrique Galvão, *Santa Maria: My Crusade for Portugal* (Cleveland, OH: World, 1961), 57, 61.

11. Nota no. 531/6ª/8, from the Senior Administrador of Cassai Sul to the *Chefes dos Postos* of Cassai Sul (October 29, 1938), 2, MAUC, Folder 86 37°.

12. Although tax rates may have been lower beyond Angola's borders, and wages, purchasing power, and even product availability higher elsewhere, residents were fundamentally fleeing the perceived harshness of mine life. For examples of colonial authorities citing tax and salary allurements as impetuses, see Alvaro Comenda, Cópia da nota no 4/R, do Administrador do Cassai-Sul (December 6, 1928), 4, AHN, Caixa 4871, Lunda; M. Lopes da Silva, "Informação" (September 26, 1941), 3, MAUC, Folder 86 40°.

13. Due to the inaccuracy of colonial censuses, it is impossible to quantify absolutely the numbers involved.

14. Miudo Rafael, interview by the author, Lunda (Angola), August 15, 2005.

15. Costa Chicungo, interview by the author, Lunda (Angola), May 13, 2005. As a result of the state's pacification campaigns and Diamang's growth, the amount of taxes collected annually in Lunda rose from 25,000 escudos in 1920 to 2,700,000 in 1937. *Boletim Geral das Colonias*, no. 162 (December 1938): 540.

16. In fact, the presence of women and children in these parties constitutes an important experiential divergence for contratados and voluntários, as the latter always proceeded to the mines unaccompanied.

17. Lisa A. Lindsay, *Working with Gender: Wage Labor and Social Change in Southwestern Nigeria* (Portsmouth, NH: Heinemann, 2003), 9. See also Frederick Cooper, *Decolonization and African Society: The Labor Question in French and British Africa* (Cambridge: Cambridge University Press, 1996).

18. Labor retention is examined in detail in chapter 7.

19. Letter from Ernesto de Vilhena to Vasco Lopes Alves (December 4, 1929), MAUC.

20. This amount was roughly equivalent to two weeks' wages for an African mine worker at Diamang.

21. MAUC, Folders 86B,6 1ª and 86 55°. From 1928 to 1958, the angolar was used as currency in Angola, which was exchanged at a ratio of 1:1 with the Portuguese escudo.

22. Diamang, Acta de 182a (March 19, 1945), 6, MAUC, Folder 86D 3°. An example from 1921 confirms that this issue was long-standing. See Letter from H. Cooper (April 26, 1921), 2, MAUC, Folder 86 1°.

23. By advocating for extended contractual periods, Diamang also hoped to reduce the costs associated with relocating workers to and from the mines and to gain access to women's labor for longer stretches.

24. Letter, Governor of Malange Province to the Colonial Overseas Minister (1952), 24, Special Collections Library at the University of Virginia (hereafter UVA).

25. Beyond the initial years of shibalo, the arrival of cipaios became part of the rhythm of rural life.

26. Janette Pedro, interview by the author, Lunda (Angola), August 16, 2005.

27. António Muiege, interview by the author, Lunda (Angola), November 21, 2005.

28. Mulevana Camachele, interview by the author, Lunda (Angola), August 16, 2005.

29. Mawassa Mwaninga, interview by the author, Lunda (Angola), November 22, 2005.

30. Fernando Tximvula, interview by the author, Lunda (Angola), August 9, 2005.

31. Mulombe Manuel, interview by the author, Lunda (Angola), August 10, 2005.

32. Letter from Borges de Sousa to Rolando de Sousa (August 19, 1954): 1, MAUC, Folder 86B, 6 7°.

33. Compiled from the following sources: *Pessoal angariado durante o ano de 1927*, MAUC, Folder 86 7°; Diamang, *Nota de informação sobre a parte relativa a mão de obra indígena do ofício dirigido* (February 22, 1938), 2, MAUC, Folder 86 35°; Direcção Técnica na Lunda, "Notas sôbre famílias de trabalhadores" (September 24, 1945), MAUC, Folder 86 46°; Letter from de Vilhena to Rolando Sucena Baptista de Sousa (February 18, 1948), MAUC, Folder 86D 6°; *Mapa de mulhers que acompanharam os contratados chegados por origens* (1952), MAUC, 86B, 6 5°; J. Robalo, *Spamoi relatório annual de 1957* (January 30, 1958), MAUC, Folder 86B, 6 9° J. Robalo, *Spamoi relatório annual de 1959* (February 16, 1960), MAUC; C. Veloso, *Spamoi relatório annual de 1960* (March 21, 1961), MAUC, 86D, 2 13°; Diamang, Direcção Técnica, Resumo dos Trabalhos Realizados em 1963 - *Relatório no. 24 (nova série), apresentado ao administrador-delegado* (October 15, 1964): 47, Arquivo Torre do Tombo (hereafter TT), Folder AOS/CO/UL-8A3, Pt. 1: 1940–64; Romano, *Relatório annual da Spamo 1964* (February 11, 1965), MAUC, Folder 86D, 2 17°; Romano, *Relatório annual da Spamo 1965* (February 8, 1966): 7, MAUC, Folder 86D, 2 18°.

34. Sacabela Sacahiavo, interview by the author, Lunda (Angola), August 12, 2005.

35. Lina Machamba, interview by the author, Lunda (Angola), August 16, 2005. Men also occasionally brought children or young relatives to the mines to assume this complementary/supportive role.

36. Interviews by the author, Lunda (Angola): Muhetxo Sapelende, August 11, 2005; Sacabela Sacahiavo, August 12, 2005.

37. Letter from A. Ferreira to Intendente do Distrito da Lunda (May 13, 1939), MAUC, Folder 86 39°.

38. Letter summarizing despacho numero 6/L, of January 5, 1922, issued by the Alto Comissario of Angola (January 16, 1922), AHN, Caixa 4871, Lunda.

39. AHN, Caixa 319, Malange, Cota #19. Much to Portugal's chagrin, University of Wisconsin Professor Edward Ross brought incidents of this

nature to international attention via his critical 1925 report, declaring: "Most of the brutality from which the natives suffer is inflicted by *cipães* who are given virtual carte blanche by their Portuguese superiors. . . . Under threat of being tied up, the villagers compete in bribing him not to hit them too hard. . . . The *cipãio* is often a criminal or a bad character" (Edward Alsworth Ross, *Report on Employment of Native Labor in Portuguese Africa* [New York: Abbott Press, 1925], 16–24). Ross did not visit Lunda while in Angola, but he did collect testimony in areas near Diamang's far western recruiting zones. Regardless, colonial and commercial officials alike generally considered the cipaio problem in Angola to be endemic.

40. Miudo Rafael, interview by the author, Lunda (Angola), August 15, 2005.

41. Lina Machamba, interview by the author, Lunda (Angola), August 16, 2005.

42. Isaacman calls for a more balanced view of these figures, drawing attention to the fact that although cipaios were scorned by their neighbors, they were also abused by European officials. Allen Isaacman, *Cotton Is the Mother of Poverty: Peasants, Work, and Rural Struggle in Colonial Mozambique, 1938–1961* (Portsmouth, NH: Heinemann, 1996), 16, 59.

43. Muhetxo Sapelende, interview by the author, Lunda (Angola), August 11, 2005.

44. António Muiege, interview by the author, Lunda (Angola), November 21, 2005.

45. Luciane Kahanga, interview by the author, Lunda (Angola), November 18, 2005.

46. For example, in 1959, the company distributed 1,025 uniforms to sobas who had "well-established histories of cooperation" and were thus "acting as 'good boys,'" as the colonial administrator of Camaxilo referred to them. Odónel Moniz, *Nota informativa sobre o pedido de fardas para sobas relativo ao ano de 1960* (December 14, 1960), MAUC, Folder 86B, 6 12°. The document also indicates that the company had distributed 899 uniforms in 1958 and approximately 1,200–1,300 in 1960.

47. Figueiredo, *Relatório Annual da Spamoi, 1962* (March 4, 1963): 40, MAUC, Folder 86D,2 15°,.

48. For an interesting examination of this practice, see Nuno Porto, "'Under the Gaze of Ancestors': Photographs and Performance in Colonial Angola," in *Photographs, Objects, Histories*, ed. Elizabeth Edwards and Janice Hart (London: Routledge, 2004), 116–24.

49. In *Terra morta*, Soromenho includes a passage that indicates that during the 1930s, the decade in which the novel is set, sobas were attempting to conceal their sons from cipaios. However, both oral and archival evidence suggest that this protective strategy was no longer possible as early as the 1940s. Castro Soromenho, *Terra morta* (Rio de Janeiro: Casa do Estudante do Brasil, 1949), 139. Diamang's director, Ernesto de Vilhena, was particularly critical of

Soromenho's accounts of rural Angola, suggesting that they were "disgusting, communist propaganda," and wondered how they survived the *Estado Novo*'s censor. Letter from Ernesto de Vilhena to António Salazar (July 29, 1958), TT, Folder 20A 2°.

50. Fina da Costa, interview by the author, Luanda (Angola), May 7, 2005. The genesis of the *imposto indígena* (indigenous tax) was the 1886 *cubata* (hut) tax, which, after an interruption from 1897 to 1907, was succeeded in 1920 by the individual imposto. Though Heywood reports that Angolan women were also liable for taxation from 1919 and were often forced onto public works projects for noncompliance, this scenario does not appear to have been the case in the areas from which Diamang traditionally drew laborers. Linda Heywood, "Slavery and Forced Labor in the Changing Political Economy of Central Angola, 1850–1949," in *The End of Slavery in Africa*, ed. Suzanne Miers and Richard Roberts (Madison: University of Wisconsin Press, 1988), 425.

51. Miudo Rafael, interview by the author, Lunda (Angola), August 15, 2005.

52. Costa Chicungo, interview by the author, Lunda (Angola), May 13, 2005.

53. A detailed examination of African laborers who completed multiple contracts appears in chapter 7.

54. João Muacasso, interview by the author, Lunda (Angola), August 11, 2005.

55. Mawassa Mwaninga, interview by the author, Lunda (Angola), November 22, 2005.

56. Mateus Nanto, interview by the author, Lunda (Angola), August 12, 2005.

57. Diamang, Acta de 182ª Sessão do Conselho (March 19, 1945), 2, MAUC, Folder 86 24°.

58. Relatório Trimestral dos Administradores (March 1952), 4, TT, Folder AOS/CO/UL-8A5.

59. Paulino, interview by the author, Lunda (Angola), August 11 and 12, 2005.

60. João Muacasso, interview by the author, Lunda (Angola), August 11, 2005. Based on Muacasso's testimony, it is possible to estimate that his party traveled hundreds of kilometers prior to reaching the posto and, factoring in spousal accompaniment rates from that year, that the group consisted of approximately 75–100 people.

61. Rodrigues, interview by the author, Lunda (Angola), August 16, 2005.

62. Muhetxo Sapelende, interview by the author, Lunda (Angola), August 11, 2005.

63. Luciane Kahanga, interview by the author, Lunda (Angola), November 18, 2005.

64. João Muacasso, interview by the author, Lunda (Angola), August 11, 2005.

65. Itela Joaquim, interview by the author, Lunda (Angola), August 12, 2005.

66. Letter from M.A.V. Osorio Junior to Rep. da Diamang (June 6, 1932), 1, MAUC, Folder 86 24°.

67. *Relatório da viagem do secretário a Vila Silva* (October 25, 1937), 13, MAUC, Folder 86 34°.

68. Letter from A. Pinto Ferreira to Sr. Agente da Diamang (May 26, 1944), MAUC, Folder 86 44°. An incident from 1941 revealed that one chefe was actually encouraging desertion. Alberto Macedo, the chefe of Lucasse posto in Moxico, allegedly told four recruits that they should "go as far as Vila Luso; after that you can flee at will." M. Lopes da Silva, "Snr. Cândido Guerreiro da Franca, Informação sobre a carta do secretário do Moxico, de September 13, 1941" (September 26, 1941), MAUC, Folder 86 40°.

69. Muatxissengue, interview by the author, Lunda (Angola), November 22, 2005.

70. Diamang, *Relatória, agencia do saurimo, serviço de recrutamento* (1931), 3, MAUC, Folder 86 23°.

71. Diamang officials were eager to both get the assembled recruits to the mines and minimize expenses, as although chefes distributed food to concentrados, these administrators in turn billed Diamang for it. Letter from José to the Chefe do Posto de Quirima (March 1, 1952), AHN, Caixa 321, Malange, Cota #1.

72. Mateus Nanto, interview by the author, Lunda (Angola), August 12, 2005.

73. Interviews by the author, Cafololo Muamuiombo and Itela Joaquim, Lunda (Angola), August 12, 2005.

74. Letter from Afonso Mendes, to Sr. Rep. da Diamang (May 1965), 1, MAUC, Folder 86B,6a 5°.

75. Fernando Tximvula, interview by the author, Lunda (Angola), August 19, 2005. For more information regarding Diamang's health practices and personnel, see Jorge Varanda, "'A Bem da Nação': Medical Science in a Diamond Company in Twentieth-Century Colonial Angola" (PhD diss., University College, London, 2007).

76. Filipe Saucauenhe, interview by the author, Lunda (Angola), November 18, 2005.

77. In 1923, a company report indicated that recruits who were carrying less than thirty kilograms would be "topped off" with company supplies, thereby turning recruits into porters. MAUC, Folder 86 2° (1923).

78. Letter from José Dias Mendes to Snr. Bernardo d'Almeida Azevedo, Chefe da Secção e Transportes, "Venda, pelos indígenas contractados, de artigos de vestuario" (May 24, 1926), MAUC, Folder 86 6°.

79. Francisco Moreira da Fonseca Abreu, *Relatório ao snr. representante da Diamang* (June 16, 1928), MAUC, Folder 86 12°. Only slightly more than half of these recruits ultimately fulfilled their contracts.

80. Letter from Antonio Brandão de Mello to the high commissioner of Angola (April 24, 1921), 5, MAUC, Folder 84K3 1°.

81. In practice, Diamang occasionally failed to sufficiently stock them. Further, during the rainy season, the shelters offered inadequate cover for large groups—or none at all if the group failed to cover the requisite distances between them.

82. Diamang, direcção técnica na Lunda (January 26, 1946), MAUC, Folder 86F, 1.

83. Although this development represented a watershed moment for contratados, the company implemented the new system rather fitfully. Moreover, the requisite bridges and rafts needed to cross Lunda's numerous rivers had to be installed, and logistical difficulties often caused significant delays at postos.

84. *Boletim Geral das Colonias*, no. 305 (November 1950): 122.

85. Dundo remains an incomparable space in Angola, with virtually all of the Diamang-era structures still intact, including rows of smart houses and lawns, as well as 1970s Volkswagen Beetles.

86. Diamang's Health Services personnel also vaccinated both incoming men and women against smallpox beginning in the late 1920s and later also treated them for yellow fever, typhoid, and intestinal parasites.

87. Letter from W. Rettie to Diamang in Luanda (June 14, 1923), 1, MAUC, Folder 84K3 2°,.

88. Letter from E. S. Lane to the Snr. Rep. da Diamang (August 21, 1928), MAUC, Folder 86 12°.

89. Maurice-Charles-Joseph, a French army doctor, created the Pignet index in 1900 as a military recruitment aid to help determine individuals' fitness levels.

90. Memorandum of Conferences held at Dundo, 1929, MAUC, Folder 86 17°.

91. In 1935, Diamang's director of engineering, Quirino Fonseca, reported that the colonial government in Katanga was using an index of "34" as its acceptance/rejection threshold, while the Union Minière de Katanga was using "30" as its cutoff. Q. Fonseca, *Relatório* (March 2, 1935), MAUC, Folder 86A7 2°.

92. 1938 company guidelines prohibited recruits with ulcers, lesions, hernias or contagious diseases. Letter from Manuel Beirão to agente de recrutamento em Vila Luso (January 15, 1939), MAUC, Folder 86 38°.

93. J. Simões Neves, *Relatório apresentados pelos administradores por parte do governo na Diamang, relativos aos anos de 1943, 1944 e 1945* (December 1946), 28, TT, Folder AOS/CO/UL-8A2.

94. Interestingly, after Diamang came under investigation by the ILO in the 1960s, the rejection figures it offered were much higher, typically around 12–13 percent. Diamang officials therefore knew that the company was contravening international standards, or at least the spirit of the labor conventions Portugal had agreed to uphold in its colonies, and thus concealed the true nature of its screening process, offering spurious figures as cover. Meanwhile, the regional African population was in many ways complicit, as, in most cases, recruits did *not* want to be rejected. See ILO Official Bulletin, # 45 (XLV) 1962, supplement II, no. 2 (April 1962): 196.

95. Domingos Cazeweque, interview by the author, Lunda (Angola), November 18, 2005.

96. Fernando Tximvula, interview by the author, Lunda (Angola), August 9, 2005.

97. "Relações de Trabalhadores Rurais Elaborados em 1963, Resumo," MAUC, Folder 86B,6a 4°; *Relatório no. 26 (nova série): Apresentado ao administrador-delegado, primeiro volume* (October 15, 1966), 60, TT, Folder AOS/CO/UL-8A3 (Cont. 2); Diamang, *Direcção técnica, resumo dos trabalhos realizados em 1965* (1966).

98. Lucas Macafuela, interview by the author, Lunda (Angola), August 16, 2005.

99. Muatxissengue, interview by the author, Lunda (Angola), November 22, 2005.

100. Because these recruits now represented significant company investments, by as early as the late 1920s Diamang was providing transport for workers between Dundo and its mines.

101. Mulevana Camachele, interview by the author, Lunda (Angola), August 16, 2005.

102. Rodrigues, interview by the author, Lunda (Angola), August 16, 2005.

103. João Muacasso, interview by the author, Lunda (Angola), August 11, 2005.

104. Luciano Xacambala, interview by the author, Lunda (Angola), November 21, 2005.

105. The conclusion of workers' contracts and their subsequent actions are examined in chapter 7.

106. With the end of shibalo, labor categories were dropped, as all employment became voluntary.

107. In rare cases, voluntários and contratados came from the same settlements. Thus, although strong correlations between provenance and engagement type existed, they were never absolute. The numbers of voluntários and contratados present on the mines over time is examined in chapter 4.

108. Costa Chicungo, interview by the author, Lunda (Angola), May 13, 2005.

109. As early as 1925, the company was issuing Diamang t-shirts to these "long-term" volunteers in the hopes that they would act as informal recruiters in their respective villages.

110. João Muacasso, interview by the author, Lunda (Angola), August 11, 2005.

111. Deque, interview by the author, Lunda (Angola), August 10, 2005.

112. Costa Chicungo, interview by the author, Lunda (Angola), May 13, 2005.

113. By the 1970s, Diamang was so appealing for regional residents that it further punished workers that had been dismissed for "grave indiscipline" by denying them reemployment during the six months that followed their termination. Reunião de chefes de grupo (April 9, 1970), 4, MAUC, Folder 86B, 6°.

114. Mulombe Manuel, interview by the author, Lunda (Angola), August 10, 2005.

115. Augusto Funete, interview by the author, Lunda (Angola), November 21, 2005.

116. Domingos Cazeweque, interview by the author, Lunda (Angola), November 18, 2005.

CHAPTER 4: A GROUP EFFORT

1. Crush, Jeeves, and Yudelman assert that "operating in the belief that this open frontier and the conditions that sustained it would never change, [South African mine] managers saw little need to improve the way they organized and used migrant labor over this long period (1897–1970)." Jonathan Crush, Alan Jeeves, and David Yudelman, *South Africa's Labor Empire: A History of Black Migrancy to the Gold Mines* (Boulder, CO: Westview Press, 1991), 3.

2. For example, on the eve of this measured push into the mechanical era, the number of *especialisados*, or skilled laborers, at Diamang stood at less than 0.5 percent (106 out of 21,632 African employees). Compagnie de Diamants de L'Angola, Reunion D'Administrateurs (October 23, 1957), 4, Museum of Anthropology, Universidade de Coimbra (hereafter MAUC), Folder 86 59°. This late push to mechanize was, in some ways, an acknowledgment that the company should have more aggressively introduced machinery earlier on, even if it might have partially undermined its paternalistic approach to labor. In addition to the expense involved, senior officials' obdurate insistence that Diamang neither needed, nor would be forced, to mechanize best explains the delay.

3. On South African mines, Moodie writes: "Management's ethnic assignations, artificial as they sometimes were, were nonetheless important in worker politics on the mines because of the way in which different types of jobs were allotted to certain groups." T. Dunbar Moodie and Vivienne Ndatshe, *Going for Gold: Men, Mines, and Migration* (Berkeley: University of California Press, 1994), 184.

4. Because the time voluntários spent in their home villages is largely inscrutable (except when bits of informants' testimony cast light onto these spaces), these laborers periodically move in and out of this chapter, as well as the ensuing ones. I have, however, attempted to minimize the disruption to the narrative as best as possible without resorting to speculation to fill these gaps.

5. Letter from G. H. Newport to direction technique (December 31, 1923), 1, MAUC, Folder 86 2°, MAUC.

6. Letter from H. T. Dickinson to direction technique (December 9, 1931), 1, MAUC, Folder 86 23°.

7. Diamang, *Memorandum on Mine Labor* (1938), 5, MAUC, Folder 86 37°.

8. Diamang, *Memorandum*, 6, MAUC, Folder 86 37°.

9. It is difficult to accurately quantify the presence of minors because Diamang often lumped them in with older workers in reports so as to avoid attracting unwanted attention from the state or, later, from outside entities critical of Portugal's colonial labor policies. Beyond the ILO, these entities included Communist governments and newly independent African states. For more information on these young workers, see Todd Cleveland, "Minors in Name Only: Child Laborers on the Diamond Mines of the *Companhia de Diamantes de Angola* (Diamang), 1917–1975," *Journal of Family History* 35, no. 1 (January 2010): 91–110.

10. In the 1960s, Diamang extended the category of minor from 16 to 18 years old.

11. In 1948, for example, minors performed 6.45 percent of all of the cascalho work (i.e., the transportation and cleaning of diamond-bearing gravel) and 13.5 percent of the prospecting duties. Mês de março, folha de distribuição do pessoal indígena homens-mês, MAUC, Folder 86 52°, 1948. Others served as messengers or "go-fers" on the mines, or even as field laborers. Informants who began at Diamang in the 1950s as minors indicated that they were mainly employed in "light" work, though their ages were lower than the company theoretically allowed. For instance, José Silva started working at Diamang in 1950 at nine years old, daily cultivating crops from 6:00 a.m. to 4:00 p.m. José Silva, interview by the author, Lunda (Angola), August 12, 2005.

12. Diamang, direcção-geral na Lunda, *Relatório no. A-549* (December 1961): 18.

13. The European labor force simultaneously increased by 48 percent due to the need for skilled laborers, so that the proportion of African workers to Europeans fell dramatically. Jean-Luc Vellut, "Mining in the Belgian Congo," in *History of Central Africa*, vol. 2, ed. David Birmingham and Phyllis M. Martin (London: Longman, 1983), 138.

14. In the 1950s, for example, Diamang created mobile work teams of between thirty and fifty laborers to address temporary shortages on individual mines.

15. For example, in 1929, Diamang designated Maludi and Luaco mines as "*voluntário*-only" assignments.

16. Letter from L. J. Parkinson to Ernesto de Vilhena (August 29, 1931), 1, MAUC, Folder 86 22°.

17. In 1930, Diamang reported that the composition of the labor force included 2,564 voluntários and 2,539 contratados, a virtually even divide and a ratio that, subsequently, was never re-approached. "Total de trabalhadores empregados na Companhia e comparação entre os voluntarios e contractados nos anos de 1927, 1928 e 1929" (1930), MAUC, Folder 86 19°.

18. Manuel P. Figueira, *Relato sôbre mão d'obra indígena* (May 18, 1938). 3, MAUC, Folder 86 36°.

19. Noronha Feyo, "Nota de Informação no. 29/46, Inspecção à mão-de-obra contratada e precauções tomadas contra acidentes de trabalho" (October 10, 1946), 1, MAUC, Folder 86A7 1°.

20. Letter from C. B. de Souza to R. Sucena de Sousa (October 22, 1948), 1, MAUC, Folder 86B, 6 1ª.

21. António Batista, interview by the author, Lunda (Angola), August 9, 2005.

22. Diamang would allocate fewer resources to prospecting when the international demand for diamonds waned, such as during the Depression, but these adjustments were both rare and temporary. Early teams operated under the threat of violence, as they were working in areas that were only nominally under colonial control, if at all. Consequently, these teams were

armed, though it appears that local communities actively challenging Portuguese encroachment directed their ire at colonial troops rather than at prospecting personnel. These teams were, however, often forced to confront sobas who insisted that they had the right to impose taxes on anyone passing through their lands, just as they had earlier on merchants. See J. Paulo, *Elementos destinados à conferência do trabalho* (June 21, 1950), MAUC, Folder 86A, 4, 1°.

23. Lute J. Parkinson, *Memoirs of African Mining* (self-published, 1962), 19.

24. Joaquim António Issuamo, interview by the author, Lunda (Angola), November 19, 2004.

25. Mário Alfredo Samuhaniquime, interview by the author, Lunda (Angola), November 18, 2005.

26. Bartolomeu Lubano, interview by the author, Lunda (Angola), August 10, 2005.

27. Caiombo Jombe, interview by the author, Lunda (Angola), August 9, 2005.

28. Mário Alfredo Samuhaniquime, interview by the author, Lunda (Angola), November 18, 2005.

29. Parkinson, *Memoirs of African Mining*, 19.

30. Diamang, *Mapa mensal de trabalhadores contratados* (July 10, 1929), MAUC, Folder 86 16°.

31. By 1948, Diamang was experiencing such difficulty in staffing prospecting teams that company officials contemplated raising the salaries of voluntários serving on prospecting teams between 30 and 60 percent before ultimately deciding against it. MAUC, Folder 86B,6 2ª, (1949–50),.

32. Pedro Bento Marques, interview by the author, Lunda (Angola), August 10, 2005.

33. Prior to the removal of overburden and cascalho, African employees cleared away any obstructive vegetation and rerouted rivers or streams if their beds were to be accessed. Whereas laborers removed vegetation manually throughout Diamang's history (subsequently using it for firewood), the company began to target the beds of fast-flowing rivers and streams beginning only in the 1960s, after it had procured enough heavy equipment to form the restraining dikes necessary to hold back the diverted flows. Physical evidence of both these procedures is still observable, and mining companies currently operating in Lunda can often determine where Diamang mined by assessing the landscape and river courses in an area. Anglophone diamond-mining staff currently working in Angola also refer to *cascalho* as "ore."

34. For two snapshots of occupational breakdowns, see Letter from H. J. Quirinho da Fonseca to Snr. Agente do Curador dos Serviçais e Colonos na Circumscription of the Fronteira do Chitato (July 28, 1931), MAUC, Folder 86 22°; Diamang, Sociedade Anónima de Responsabilidade Limitada, *Relatório do conselho de administração e parecer do conselho fiscal* (Lisbon: Diamang, 1964).

35. Letter from Ernesto de Vilhena to W. A. Odgers (October 20, 1947), MAUC, Folder 86 51°.

36. Letter from W. A. Odgers, to Ernesto de Vilhena (September 26, 1947), 1, MAUC, Folder 86 51°.

37. In order to maximize profitability, at least until the 1960s when heavy machinery was more prominent, Diamang focused on areas that featured fewer than ten meters of overburden.

38. Mualesso Gaston, interview by the author, Lunda (Angola), August 9, 2005.

39. J. Simões Neves, *Relatório apresentados pelos administradores por parte do governo na Diamang, 1941 e 1942* (October 1943), 10, Arquivo Torre do Tombo (hereafter TT), Folder AOS/CO/UL-8A3, Pt. 1: 1940–64.

40. Muhetxo Sapelende, interview by the author, Lunda (Angola), August 11, 2005.

41. Although this technology had been in use in mines elsewhere from the middle of the nineteenth century, the company was able to introduce it only fitfully, struggling to construct its own monitors in the field.

42. *Notas sobre a emprêgo das pás mecânicas "Ruston-Bucyrus" no desmonte e remoção de estéril* (May 14, 1949), 13, MAUC, Folder 86 55°. By 1952, one half of the overburden removed on the company's mines was done via hydraulic process. Diamang, Sociedade Anónima de Responsibilidade Limitada, *Relatório de conselho de administração e parecer do conselho fiscal* (Lisbon: Diamang, 1953).

43. Diamang, Sociedade Anónima de Responsibilidade Limitada, *Relatório de conselho de administração e parecer do conselho fiscal relativos ao exercício de 1950* (Lisbon: Diamang, 1951); Diamang, Sociedade Anónima de Responsabilidade Limitada, *Relatório do conselho de administração e parecer do conselho fiscal relativos ao exercício de 1960* (Lisbon: Diamang, 1961).

44. Luciane Kahanga, interview by the author, Lunda (Angola), November 18, 2005.

45. Rodrigues, interview by the author, Lunda (Angola), August 16, 2005.

46. According to mine inspector António Botelho, at least some production loss was supposedly inevitable because "Africans are more susceptible to cold than heat and are particularly affected by the rains. It takes them some time after a good rain to return to their normal rates of productivity, it promotes absenteeism and even disorients those who arrive late, claiming it skews their sense of time." António Botelho, "Inspecção do rendimento da mão-de-obra, nota de informação no. 6/70, algumas considerações sobre produtividade" (January 31, 1970), 3, MAUC, Folder 11-C (1970).

47. Mualesso Gaston, interview by the author, Lunda (Angola), August 9, 2005.

48. Paul Decauville (1846–1922), a French pioneer in light railways, invented this system in the second half of the nineteenth century. His major innovation was the use of ready-made sections of light, narrow-gauge track fastened to steel sleepers. This track was portable and could easily be disassembled and transported.

49. By 1947, for example, Diamang had 1,682 trams and over 110 kilometers of track in service.

50. Dinis dos Santos Muriandambo, interview by the author, Lunda (Angola), August 11, 2005.

51. Luciane Kahanga, interview by the author, Lunda (Angola), November 18, 2005.

52. Lutero Almeida, "Nota de informação no. 111/70" (October 6, 1970), 4. MAUC, Folder 84G, 1 3.

53. António Ramos, "Resposta" (December 12, 1970), MAUC, Folder 84G, 1 3.

54. A March 1948 report confirms that voluntários performed 85 percent of the Central Selection Station work. *Folha de distribuição do pessoal indígena homens-mês* (March 1948), MAUC, Folder 86 52°.

55. These devices were known as Joplin jigs due to their origination on the zinc mines of Joplin, Missouri. Parkinson described the process as follows: "A Joplin jig rested on a fulcrum and was lowered into a box filled with water, where, by means of an up and down motion imparted at the far end by the native, the heavier material (diamonds, etc.) settled gradually to the bottom." Parkinson, *Memoirs of African Mining*, 26.

56. Until 1923, pans were all hand operated, before steam-powered models completely replaced them by 1928. Diamang initially situated pans near excavation sites, but as transport improved, it placed them in centrally located lavarias, where workers fed them with cascalho arriving from multiple mines. Over time, as Diamang increased its reliance on pans, equipment breakdowns hampered production more dramatically. In the beginning of 1950, for example, the company reported that "work in four mines has been suspended and activity reduced in three others due to the 'paralyzation' of one of the two lavarias that service these two mine clusters." Diamang, Sociedade Anónima de Responsibilidade Limitada, *Relatório de conselho de administração* (Lisbon: Diamang, 1951), MAUC, Folder 86 53°.

57. Technicians from South Africa had originally arrived in 1950 to assist the company with the introduction of this machinery, but the lack of key parts and knowledge delayed its introduction for another five years!

58. Diamang adopted this load sharing strategy going forward. For example, in 1969, there were forty-nine lavarias and three MD machines in use.

59. It appears that at this point in time Diamang was gearing up to treat a series of potentially lucrative kimberlite deposits that prospectors had identified. Bernardo Reis, ed., *Diamante* 22 (Braga, 2004): 55.

60. Silvestre Muachembe, interview by the author, Lunda (Angola) November 19, 2004.

61. Parkinson, *Memoirs of African Mining*, 27.

62. Paulo Leão Vega, interview by the author, Lunda (Angola), May 17, 2005.

63. Leonardo Manuel Judas Chagas, interview by the author, Lisbon, June 7, 2006.

64. Carlos Machado, interview by the author, Lisbon, March 18, 2006.

65. While many informants confirmed that Chagas's first claim was accurate, I was unable to locate any former pickers who may have been able to confirm or deny the second one.

66. Leonardo Manuel Judas Chagas, interview by the author, Lisbon, June 7, 2006.

67. Apparently, this problem was most acute on cloudy days when the sun could not be relied on as a natural rouser. Letter from João Augusto Bexiga (July 21, 1971), 2, MAUC, Folder 11-C (1971).

68. Letter from João Augusto Bexiga, *Notas de informação nos. 43 e 44/71 2* (1971), MAUC, Folder 11-C. For an interesting exploration of African workers and "Western" labor schedules, see Keletso Atkins, *The Moon Is Dead! Give Us Our Money!: The Cultural Origins of an African Work Ethic, Natal, South Africa, 1843–1900* (Portsmouth, NH: Heinemann, 1993).

69. Noronha Feyo, "Nota de informação no. 29/46" (October 10, 1946), 2, MAUC, Folder 86A7 1°.

70. Unlike in diamond mines in South Africa or Southwest Africa (Namibia), Diamang's African laborers did not have to countenance invasive searches or, later, x-rays as they passed into and out of the mining sites, as the Angolan enterprise deemed its antitheft measures to be sufficient. See, for example, Robert J. Gordon, *Mines, Masters and Migrants: Life in a Namibian Compound* (Johannesburg: Ravan Press, 1977), and William H. Worger, *South Africa's City of Diamonds: Mine Workers and Monopoly Capitalism in Kimberley, 1867–1895* (New Haven, CT: Yale University Press, 1987).

71. Caiombo Jombe, interview by the author, Lunda (Angola), August 9, 2005.

72. Inspection reports of Catongula and Cassiaxima mines from 1946 indicated that cafeterias were "still absent" from these sites, suggesting that this deficiency was noteworthy. Diamang, Inspecção dos Serviços de Administração (May 6, 1946), MAUC, Folder 86 47°; Noronha Feyo, "Inspecção dos serviços, nota de informação no. 18/46" (August 8, 1946), 3, MAUC, Folder 86D 4°.

73. Letter from W. H. Newport to Mr. Shaler (April 16, 1924), 6, MAUC, Folder 86 3°.

74. Charles van Onselen, *Chibaro: African Mine Labour in Southern Rhodesia, 1900–1933* (London: Pluto Press, 1976), 51. Although van Onselen's study ends in 1933, at a point when Diamang was itself just achieving food security, even prior to that time undernourishment on Angola's mines was not nearly as severe.

75. Janette Pedro, interview by the author, Lunda (Angola), August 16, 2005.

76. Mulevana Camachele, interview by the author, Lunda (Angola), August 16, 2005.

77. Colonial legislation enacted in 1922 and 1928 formalized this work schedule, but by the latter date Diamang had already been operating in this manner for almost a decade.

78. The company had experimented with variations of this system earlier, motivated by the notion that strong and weak workers were capable of reaching different daily targets, but these efforts were neither widespread nor sustained. See, for example, Letter from S. T. Kelsey to Snr. Chefe da Circunscrição de Fronteira do Chitato (November 25, 1925), 2, MAUC, Folder 86 4°.

79. Diamang also instituted the task system to counter work "slowdowns," in which laborers deliberately gave less than their all and thereby hindered production (examined in chapter 5), though these were exceedingly rare, especially following the emerging professionalism of the African labor force.

80. Noronha Feyo, "Inspecção dos serviços, nota de informação no. 4/47" (February 19, 1947), MAUC, Folder 86D 5°. Feyo still noted that the mine boss was "kindhearted" and that the attitude among his employees was excellent. A few informants also commented that overseers physically abused workers who failed to finish their tasks, though most mine bosses and capitas recognized that this type of response was ultimately counterproductive and was, therefore, rare. Mualesso Gaston, interview by the author, August 9, 2005.

81. Mulombe Manuel, interview by the author, Lunda (Angola), August 10, 2005.

82. Noronha Feyo, "Nota de informação no. 17/46" (August 29, 1946), 2, MAUC, Folder 86D 4°.

83. Parkinson, *Memoirs of African Mining*, 39.

84. The dismissal in 1930 of a particularly violent Russian overseer named Glouschkoff, whose reputation was allegedly scaring away potential volunteers, constitutes the only instance of dismissal during this early period of operations. "Dismissal of the Agent Glouschkoff" (September 9, 1930), MAUC, Folder 86 21°.

85. Letter from F. Martins to the Director Geral da Diamang (February 3, 1947), 1, MAUC, Folder 86 49°.

86. Letter from Fernando Barros Xavier Martins, O Agente de Curador, to the Director Geral da Companhia de Diamantes de Angola (October 28, 1947), 1, MAUC, Folder 86 51°,.

87. Noronha Feyo, "Inspecção dos serviços de administração, nota de informação no. 20/46" (September 10, 1946), MAUC, Folder 86D 4°.

88. According to Ball, the Cassequel Company significantly improved conditions following the outbreak of the War for Independence. Similarly, Penvenne indicates that in Mozambique "forced labor sentences and *palmatória* [a type of paddle struck against recipients' palms] beatings peaked with the labor shortage in the early 1950s, and declined rapidly after 1957. After Portugal signed the ILO statutes and moved toward further integration with Europe, fines became the most common punishment for all offenders." Jeremy Ball, "'The Colossal Lie': The Sociedad Agrícola do Cassequel and Portuguese Colonial Labor Policy in Angola, 1899–1977" (PhD diss., UCLA, 2003); Jeanne Marie Penvenne, *African Workers and Colonial Racism: Mozambican Strategies and Struggles in Lourenço Marques, 1877–1962* (Portsmouth, NH: Heinemann, 1995), 115.

89. Ernesto de Vilhena, *Aventura e rotina: Crítica de uma crítica* (Luanda: Emprêsa Gráfica, 1954), 21. Parkinson echoed, "If such paternalism is a crime, then let there be more crime!" Parkinson, *Memoirs of African Mining*, 70.

90. Letter from A. Mendes, to the Rep. da Diamang (Jan. 28, 1965), 1, MAUC, Folder 86B, 6a 5°.

91. Mulombe Manuel, interview, Lunda (Angola), August 10, 2005; Joaquim Muamungo, interview by the author, Lunda (Angola), November 18, 2005.

92. Dinis dos Santos Muriandambo, interview by the author, Lunda (Angola), August 11, 2005. Although most corporal punishment that capitas delivered had been ordered by white supervisors, these African overseers also occasionally tormented workers on their own volition. In 1946, Diamang concluded that some capitas were guilty of appropriating portions of workers' daily rations, and in 1955, it accused a group of eleven capitas of similar improprieties. A. Borges, Direcção Técnica na Lunda, Secção de Trabalho Indígena, "Nota de informação no. 5/46, Assunto: Gratificações em géneros a trabalhadores voluntários" (March 27, 1946), MAUC, Folder 86 47°.

93. Caiombo Jombe, interview by the author, Lunda (Angola), August 9, 2005.

94. Domingos Matos, interview by the author, Lunda (Angola), November 18, 2005.

95. Muatxinjango Maca, interview by the author, Lunda (Angola), November 21, 2005.

96. Parkinson, *Memoirs of African Mining*, 93.

97. For an example of Namibian diamond mine workers being denied pensions, see Gordon, *Mines, Masters and Migrants*, 77.

98. Letter from H. J. Q. Fonseca to Snr. Agente do Curador dos Serviçais e Colonos na Circunscrição de F. do Chitato, "Acidentes de trabalho" (June 28, 1929), MAUC, Folder 86 16°.

99. Oliveira, "Relatório" (October 28, 1929), Folder 86 17°, MAUC; (December 18, 1929), MAUC, Folder 86 16°.

100. *Relatório apresentados pelos administradores por parte do govérno na Companhia de Diamantes de Angola relativos ao ano de 1938* (March 29, 1940), 18, TT, Folder AOS/CO/UL-8A3, Pt. 1: 1940–64.

101. Deque, interview by the author, Lunda (Angola), August 10, 2005.

102. João Paulo Sueno, interview by the author, Lunda (Angola), August 12 and 15, 2005. Many minors were able to utilize the language skills they had acquired while working in Portuguese homes to act as translators between African overseers and mine bosses. Gabriel Alberto, interview by the author, Lunda (Angola), August 16, 2005.

103. Noronha Feyo "Inspecção dos serviços de administração, nota de informação no. 37/46" (December 30, 1946), MAUC, Folder 86D 5°.

104. A 1971 manual for mine bosses instructed them not to alter the scene of an accident so that investigators could properly determine how and why it happened, which implies that the company was more interested in ensuring that accidents didn't recur, rather than covering up these incidents. António Botelho, *Diamang,Guia do empregado da mina* (Diamang: 1971), 73.

105. Vasco José de Oliveira, *Relatório* (October 28, 1929), 1, MAUC, Folder 86 17°.

106. Ibid. Oliveira did concede that the company may see fit to offer an indemnification anyway, due to its "fairness," but he was certain that the Diamang representative "will agree with my handling [of the case]."

107. Mário Alfredo Samuhaniquime, interview by the author, Lunda (Angola), November 18, 2005.

108. Originally, only contratados could access these posts, but this policy proved impractical because the posts were primarily intended to address medical issues (injuries) that required immediate attention, and thus over time Diamang extended access to voluntários as well.

109. Felipe Leo Muatxissupa, interview by the author, Lunda (Angola), August 15, 2005.

110. Costa Chicungo, interview by the author, Lunda (Angola), May 13, 2005.

111. Alberto Rossa, interview by the author, Lunda (Angola), November 16, 2004.

112. Letter to the governor general of Angola (1952), 7, Special Collections Library at the University of Virginia (hereafter UVA).

113. Peter Carstens, *In the Company of Diamonds: De Beers, Kleinzee, and the Control of a Town* (Athens: Ohio University Press, 2001), 185.

114. Letter to Glenn H. Newport from E. Torre do Valle (October 25, 1925), 4, MAUC, Folder 86 4°.

115. Letter from Eduardo de Vilhena to António Salazar (March 8, 1961), TT, Folder 20A 4°.

116. Joaquim Ezaia, interview by the author, Lunda (Angola), August 10 and 11, 2005.

117. Joaquim Trinidade, interview by the author, Lunda (Angola), November 15, 2004.

118. The inclusion of the many voluntário women who served as nurses or health auxiliaries would increase this figure slightly, but as they served mainly in Dundo, far from the mines, they remain outside the focus of this study.

119. With this initiative, the company was proactively displaying a degree of magnanimity, as even the CTI of 1929 required that women be paid only half of a male employee's salary. Julio Callaça, *Código do trabalho dos indígenas* (October 1, 1929), MAUC, Folder 86 18°.

120. Janette Pedro, interview by the author, Lunda (Angola), August 16, 2005.

121. Mulevana Camachele, interview by the author, Lunda (Angola), August 16, 2005.

122. Lina Machamba, interview by the author, Lunda (Angola), August 16, 2005. *Funge*, made from ground manioc or, less commonly, ground maize, is an Angolan staple.

123. Lina Machamba, interview by the author, Lunda (Angola), August 16, 2005. Machamba indicated that at her first kitchen there were approximately ten staff members, serving a mine population of roughly two hundred people.

124. This arrangement resembled scenarios in Northern Rhodesia in which wives accompanied their husbands to copper mines and freely cultivated land that mining companies made available to them. In 1925, Diamang also began to supply a half ration to the few women who worked on the

company's incipient plantations. See George Chauncey Jr., "The Locus of Reproduction: Women's Labour in the Zambian Copperbelt, 1927–1953," *Journal of Southern African Studies* 7, no. 2 (1981): 138–39.

125. Lina Machamba, interview by the author, Lunda (Angola), August 16, 2005.

126. H. J. Quirino Fonseca, "Situation of the Fuba Market" (November 8, 1938), 5, MAUC, Folder 86 37°.

127. *SPAMOI Relatório Annual de 1953* (January 27, 1954), MAUC, Folder 86D, 2 7°.

128. Sacabela Sacahiavo, interview by the author, Lunda (Angola), August 12, 2005.

129. Mawassa Mwaninga, interview by the author, Lunda (Angola), November 22, 2005.

130. *Spamoi report* (June 1941), 8, MAUC, Folder 86D 2°.

131. Janette Pedro, interview by the author, Lunda (Angola), August 16, 2005.

132. J. Robalo, *Spamoi report* (September 8, 1945), MAUC, Folder 86D 4° .

133. *Spamoi Relatório Anual de 1952* (February 14, 1953), 5, MAUC, Folder 86D, 2 6°; Canhão Veloso, *SPAMOI Relatório Annual de 1953* (January 27, 1954), 5, MAUC, Folder 86D, 2 7°.

134. J. Robalo, "Namutondo, wife of contratado, Capumba" (February 28, 1959), MAUC, Folder 86D 6°.

135. Mário Correia, "Relatório" (March 21, 1959), MAUC, Folder 86D 6°.

136. Acts of this nature violated Service Order 11-D/48 of May 31, 1948, which called for guilty parties to be "punished with the most severe sanctions, including even termination of employment."

137. Letter from J. Tavares Paulo to the Chefe da Spamoi (April 17, 1950), 1, MAUC, Folder 86B, 6 3°.

138. Letter from J. Tavares Paulo to the Chefe da Spamoi (April 17, 1950), 1, MAUC, Folder 86B, 6 3°.

139. Janette Pedro, interview by the author, Lunda (Angola), August 16, 2005.

140. Anna Maria Dos Santos, interview by the author, Lunda (Angola), August 10, 2005.

CHAPTER 5: NEGOTIATING STABILITY

1. Alberto Rossa, interview by the author, Lunda (Angola), November 16, 2004.

2. António Batista, interview by the author, Lunda (Angola), August 9, 2005. The company sometimes referred to task-based workers as *empreiteiros*, or "contractors," which is how Batista identified himself.

3. Sacabela Sacahiavo, interview by the author, Lunda (Angola), August 12, 2005. Another informant, Mualesso Gaston, indicated that workers would be made to finish the uncompleted work of any laborers hauled off for punishment, suggesting that helping out was a preemptive endeavor that may

otherwise have become mandatory, anyway. Mualesso Gaston, interview by the author, Lunda (Angola), August 9, 2005.

4. Mulombe Manuel, interview by the author, Lunda (Angola), August 10, 2005.

5. Daniel Mendes and Romano, *Spamoi relatório no. 10/62 de Outubro 1962* (November 20, 1962), 9, Museum of Anthropology, Universidade de Coimbra (hereafter MAUC), Folder 86D, 2 15°.

6. Janette Pedro, interview by the author, Lunda (Angola), August 16, 2005.

7. Mulevana Camachele, interview by the author, Lunda (Angola), August 16, 2005.

8. Lute J. Parkinson, *Memoirs of African Mining* (self-published, 1962), 52.

9. Bartolomeu Lubano, interview by the author, Lunda (Angola), August 10, 2005. Other informants reported that the company employed an African overseer (*capita*) to interact with workers in a "call and response" fashion to keep them animated while working.

10. Costa Chicungo, interview by the author, Lunda (Angola), May 13, 2005.

11. Interviews by the author, Paulo Leão Vega and António Sulessa, Lunda (Angola), May 17, 2005.

12. João Muacasso, interview by the author, Lunda (Angola), August 11, 2005.

13. Costa Chicungo, interview by the author, Lunda (Angola), May 13, 2005.

14. Mawassa Mwaninga, interview by the author, Lunda (Angola), November 22, 2005.

15. Noronha Feyo, "Nota de informação no. 18/46" (August 8, 1946), 3, MAUC, Folder 86D 4°.

16. Noronha Feyo, "Nota de informação no. 17/46" (August 29, 1946), 3, MAUC, Folder 86D 4°.

17. Noronha Feyo, "Nota de informação no. 29/46, inspecção à mão-de-obra contratada e precauções tomadas contra acidentes de trabalho" (October 10, 1946), 3, MAUC, Folder 86A7 1°. *Preto* is a pejorative term in Portuguese for a black African.

18. Noronha Feyo, "Nota de informação no. 25/46" (September 17, 1946), MAUC, Folder 86D 4°.

19. Noronha Feyo, "Nota de informação no. 25/46" (September 17, 1946), MAUC, Folder 86D 4°.

20. For example, Gordon states that on Namibia's diamond mines, "African workers felt it was useless to complain about conditions." Robert J. Gordon, *Mines, Masters and Migrants: Life in a Namibian Compound* (Johannesburg: Ravan Press, 1977), 93.

21. Letter from H. T. Dickinson, consulting engineer, to Bureau Technique, Bruxelles, "Visit of Governor Roma" (August 21, 1925), 5–6, MAUC, Folder 86 3°.

22. Letter from E. Torre do Valle to Glenn H. Newport (October 25, 1925), 1, MAUC, Folder 86 4°.

23. Women, too, often complained about workplace conditions. For example, Namutondo, who had been a victim of abuse at the hands of A. Lopes (as outlined in the previous chapter), was, in fact, also the person who brought this episode to the company's attention via a formal complaint. "Namutondo, wife of contracted worker no. 4698, Capumba" (February 28, 1959), MAUC, Folder 86D 6°.

24. Letter from Victor Hugo Antunes to Snr. Engenheiro da Diamang (November 16, 1927), 1, MAUC, Folder 86 8°.

25. Letter from Governador Bento Roma to Snr. Delegado do Representante da Diamang (December 15, 1925), 1, MAUC, Folder 86 4°.

26. MAUC, Folder 86 8° (1928).

27. Costa Chicungo, interview by the author, Lunda (Angola), May 13, 2005.

28. Romeu Moura, "Copia, auto de declarações" (August 5, 1930), 1–2, MAUC, Folder 86 21°.

29. Alberto Rossa, interview by the author, Lunda (Angola), November 16, 2004.

30. Anna Maria Dos Santos, interview by the author, Lunda (Angola), August 10, 2005.

31. Felipe Leo Muatxissupa, interview by the author, Lunda (Angola), August 15, 2005.

32. Diamang's labor department's 1972 annual report declared that owing to "increased literacy . . . the number of letters submitted by workers rose extraordinarily, presenting the most varied complaints. These were always promptly attended to, when pertinent. If they lacked any basis, we explained to the workers—in writing or in person—of their lack of reason." Yet even at this late date, only a handful of African employees were literate, so the number of written complaints was surely small, even if increased. See Secção de expediente da mão-de-obra, *Relatório annual* (1972), 2, Arquivo Torre do Tombo (hereafter TT), Folder 86C, 2.

33. This reactive measure also effectively capped laborers' daily effort level by allowing them to work only as hard as was necessary to reach their assigned target, while capitas could do little to further motivate workers as long as they were meeting these daily task quotas.

34. Unlike van Onselen, I do not consider that the objective of "loafing" was to "steal" from the employer, nor do I treat these acts as a "conscious rebellion against the employer," but rather as strategies intended to immediately attenuate the daily labor regime. Charles van Onselen, *Chibaro: African Mine Labour in Southern Rhodesia, 1900–1933* (London: Pluto Press, 1976), 241.

35. Letter from E. S. Lane to Snr. Chefe da Circumscrição de Fronteira do Chitato, "Indígenas chitambala e sachimboio" (December 12, 1927), MAUC, Folder 86 9°.

36. Letter from E. S. Lane, representante na Lunda, to Snr. Chefe da C de Fronteira do Chitato, "Ocorrencias com trabalhadores nas explorações" (September 12, 1928), 1–3, MAUC, Folder 86 13°.

37. Letter from Lute J. Parkinson to E. S. Lane (December 21, 1928), MAUC, Folder 86 14°.

38. Costa Chicungo, interview by the author, Lunda (Angola), May 13, 2005.

39. Luciane Kahanga, interview by the author, November 18, 2005. Kahanga explained that capitas attempted to keep workers on pace by barking out a rhythmic "4 by 4" cadence.

40. As most voluntários were paid only if they worked, this strategy was primarily employed by contratados.

41. A. A. de Almeida e Souza, *Relatório medico sobre movimento hospitalar dos trabalhadores indígenas* (September 20, 1930), 3, MAUC, Folder 86 21°.

42. A. A. de Almeida e Souza, *Relatório medico*, 2, MAUC, Folder 86 21°.

43. Letter from Ernesto de Vilhena to Vasco Alves (December 14, 1929), 6, MAUC, Folder 86 18°.

44. For example, see J. Simões Neves, *Relatorio apresentado pelos administradores por parte do governo na Companhia de Diamantes de Angola eelativo ao ano de 1936* (December 1937), 17, TT, Folder AOS/CO/UL-8A; J. Simões Neves, *Relatório* (December 1946), 26, TT, Folder AOS/CO/UL-8A2.

45. Ideally, medical personnel made this assessment, but at other times (unqualified) mine bosses sufficed.

46. Vasco José de Oliveira, *Relatório* (October 28, 1929), MAUC, Folder 86 17°.

47. António Botelho, *Diamang, guia do empregado da mina* (Diamang, 1971), 73.

48. Bernardo Montaubuleno, interview by the author, Lunda (Angola), May 18, 2005.

49. Unsigned letter attached to Letter from Victor Hugo Antunes, o chefe da circunscrição (fronteira do chitato), to Snr. Engenheiro Sub-Director da Diamang (November 16, 1927), MAUC, Folder 86 8°.

50. Alberto Rossa, interview by the author, Lunda (Angola), November 16, 2004. Rossa added that, at times, a lighter work load or task may also have acted as a laborer's motivation to feign illness or injury.

51. Bernardo Montaubuleno, interview by the author, Lunda (Angola), November 20, 2004.

52. Rather than considering theft as a wage supplement, linked to low rates of pay, as Cohen has argued, or as an appropriation of an item that workers considered to be "owed to them," as Scott has contended, I understand this strategy to have been an endeavor in which workers engaged simply to capitalize financially on their time on the mines, free from these types of justifications or rationalizations. Robin Cohen, "Resistance and Hidden Forms of Consciousness amongst African Workers," *Review of African Political Economy* 7, no. 19 (September–December 1980): 20; James C. Scott, *Weapons of the Weak: Everyday Forms of Peasant Resistance* (New Haven, CT: Yale University Press, 1985), 269.

53. This sentiment is extremely surprising given Diamang officials' awareness of theft on both Forminière's mines and South Africa's array of diamond mines.

54. Parkinson, *Memoirs of African Mining*, 25.

55. Letter from H. T. Dickinson to Ernesto de Vilhena (January 14, 1926), 20, MAUC, Folder 86 5°. Despite Dickinson's astuteness, he also mistakenly surmised in the same letter, "It is clear that once we can prevent the natives employed in the picking of stones from appropriating them on their own behalf, the present illicit diamond trade will be almost entirely suppressed." He also, however, correctly implicated whites in the trade: "But, vigilance should also be exercised over white personnel, as there are certain cases of theft which can only be accounted for by the intervention of some of the white employees."

56. Unfortunately, it is impossible to determine how many stones left Diamang's mines undetected or even how much these audacious, frontline African operators were earning for their efforts. As part of Decree no. 12:148 of August 19, 1926, Portugal's Minister of the Colonies declared: "When the possibilities for the production of diamonds on a large scale in Angola were not well known on the international markets, the illegal exit of diamonds from its mines was nil, or insignificant; but as soon as these possibilities and the good quality of its stones became duly appreciated, the colony became the object of attempts analogous to those in other countries [the Belgian Congo and South Africa] that have been victimized with such great success that it was obvious that, related to this type of criminal activity, we lacked not only an effective penal repression system, but even a satisfactory police organization."

57. Diamang, *Legislação penal referente à exploração de diamantes na colónia de Angola* (Lisbon: Oficina Gráfica, 1948), 4.

58. In the 1920s, Diamang officials suspected that Africans were enticed using gunpowder, gun caps, or fuses, the sale of which was prohibited in the colony.

59. MAUC, Folder 86 11° (1928). Whites in the area were originally allowed to possess rough diamonds, but after 1925, at the urging of Diamang, the colonial state expressly forbade private ownership.

60. Letter from Osório Júnior to the Snr. Administrador da CC do Cassai Sul, "Trabalhadores" (July 11, 1936): 1, MAUC, Folder 86 31°.

61. This figure represented only 0.63 percent of the total production (1,298,038 carats) to date. MAUC, Folder 86 32° (1937). As Carruthers's work on Africans "poaching" in colonial game parks suggests, Diamang didn't necessarily rue the actual (minute) loss as much as this expression of freedom. Jane Carruthers, "Creating a National Park, 1910 to 1926," *Journal of Southern African Studies* 15, no. 2 (1989): 201.

62. A. Metello, "A repressão do transito clandestino" (October 10, 1931), 7, MAUC, Folder 86 23°.

63. By this time, Diamang claimed to have recovered some 9,189 carats.

64. Letter from Ernesto de Vilhena to Sr. Dr. Eugenio Ferreira, "Expulsão de indígenas para São Tomé" (February 5, 1947), MAUC, Folder 86 49°.

65. Cópia de telegrama informando a descoberta de autores de assalto, sent from Vila Henrique Carvalho to Lisbon (January 3, 1946), TT, Folder AMC8, 8. In all, ten stones were recovered, though no carat count was recorded. For his crimes, Sucasuca received an unprecedented fifteen-year jail

term, which he served at Angola's Colónia Penal Agrícola da Damba. An interesting company letter from 1961 documents his release and, given his record, warns against satisfying his request to relocate. Letter from Gil Germano Gonçalves Ferreira to the P. A. de Mucári, "No 1561/6ᵃ/24, Transcription of nota no. 2118/2ᵃ/9 of May 1961" (1961), Arquivo Histórico Nacional de Angola (hereafter AHN), Caixa 4410, "Malange."

66. Diamang, Sociedade Anónima de Responsabilidade Limitada, *Relatório do conselho de administração e parecer do conselho fiscal relativos ao exercício de 1957 para serem presentes á assembleia geral ordinária de 30 de Junho de 1958* (Lisbon: Diamang, 1958), 97.

67. Diamang, *Informação, assunto: Furto de um diamante* (1969), MAUC, Folder 86B, 6°.

68. Ibid. These comments are somewhat surprising, as following the outbreak of the Angolan revolution in 1961 both the state and company were concerned that diamonds smuggled out of Lunda might be used to bankroll the various independence movements' activities.

69. Leonardo Manuel Judas Chagas, interview by the author, Lisbon, June 7, 2006.

70. Paulo Leão Vega, interview by the author, Lunda (Angola), May 17, 2005. In *Terra Morta*, Soromenho writes that a diamond was often referred to as a *"feijão branco"* (white bean), presumably for the same reason as Vega cited. Castro Soromenho, *Terra morta* (Rio de Janeiro: Casa do Estudante do Brasil, 1949), 76.

71. Paulo Leão Vega, interview by the author, Lunda (Angola), May 17, 2005.

72. Andre Muamukepe, interview by the author, Lunda (Angola), May 18, 2005.

73. Isabel Reis, interview by the author, Oeiras (Portugal), June 15, 2004.

74. Letter from G. H. Newport to Diamang in Brussels (December 31, 1923), 1–2, MAUC, Folder 86 2°.

75. Letter from E. S. Lane to the Snr. Secretario do Distrito (March 16, 1929), 1, MAUC, Folder 86 15°.

76. Letter from H. T. Dickinson to M. K. Shaler (February 14, 1931), 4, MAUC, Folder 86 22°.

77. Letter from Lute J. Parkinson to Ernesto de Vilhena (August 29, 1931), 1, MAUC, Folder 86 22°.

78. Memorandum sobre mão d'obra indígena (May 27, 1934), 9, MAUC, Folder 86 26°; *Extracto de uma nota do eng. Quirinho da Fonseca* (1935), 2, MAUC, Folder 86 29°. The exact figures for voluntários were: 1930: 54.96 percent; 1931: 64.38 percent; 1932: 67.50 percent; 1933: 70.69 percent; 1934: 68.82 percent.

79. Letter from H. J. Quirino Fonseca to unknown recipient (July 21, 1931), MAUC, Folder 86 22°.

80. *Relatório apresentado pelos administradores por parte do governo na companhia de diamantes de Angola relativo ao ano de 1935* (December 31, 1936), 11, TT, Folder AOS/CO/UL-8A, TT; *Relatório de Julho de 1935, secção de trabalho indígena* (August 9, 1935), MAUC, Folder 86 28°, MAUC.

81. Spamoi report (June 1941), 1, MAUC, Folder 86D 2°, MAUC; Diamang, Direcção-Geral na Lunda, *Relatório no. A-549* (December, 1961), 8.

82. MAUC, Folder 86 32°,(1937).

83. Spamoi report (June 1941), 4, MAUC, Folder 86D 2°.

84. Santos Ribeiro, "Nota de informação" (August 19, 1949), 1, MAUC, Folder 86A7 3°. According to Diamang officials, ethnic Camatapas from near Canzar posto were notorious for their absenteeism.

85. Anna Maria dos Santos, interview by the author, Lunda (Angola), August 10, 2005.

86. "Nota de informação no. 23/72, inspecção da mina Saga" (1972), MAUC, Folder 86A7 31°.

87. António Botelho, "Inspecção do rendimento da mão-de-obra, nota de informação no. 6/70, algumas considerações sobre produtividade" (January 31, 1970), 13, MAUC, Folder 11-C (1970).

88. Letter from João Augusto Bexiga, o director geral na Lunda, to administração da Diamang, Lisboa, "Notas de informação nos. 43 e 44/71" (July 21, 1971), 3, MAUC, Folder 11-C (1971).

89. In an attempt to lower absenteeism rates, Diamang periodically experimented with rewards for steady attendance. Initially, enticements consisted of small bean or fish allotments, though these met with only limited success. In July 1970, the company concluded that beer would serve as an effective enticement. Workers welcomed this initiative, though they failed to pursue absenteeism with any lesser—or greater—vigor. Alternatively, Diamang began awarding small monetary bonuses to entire workforces on mines that featured absenteeism rates lower than 2 percent for a series of months. Other rewards were based on mine workers' specific requests, which resulted in the introduction of recreation centers on two different lavarias in 1972. "Nota de Informação No. 176/72" (1972), MAUC, Folder 86A, 7 32.

90. In 1931, for example, Diamang reported that a number of workers were fleeing to Moxico Province (south of Lunda), aware that the company was not (yet) recruiting there. Diamang, *Relatório, agencia do saurimo, serviço de recrutamento* (1931), 3, MAUC, Folder 86 23°.

91. Letter from the administrador (October 15, 1925), 1, AHN, Caixa 2084 "Cachingues."

92. Letter from H. T. Dickinson to Direction Technique (November 4, 1931), 1, MAUC, Folder 86 23°.

93. Letter from E. Machado to Direcção Tecnica da Lunda (February 26, 1929), MAUC, Folder 86 19°.

94. In turn, Diamang begrudgingly acknowledged that there was, indeed, a bronchial-pneumonia flu that was afflicting a number of workers. Letter from José Maria de Noronha Feyo, o agente em Vila Luzo, to Snr. Governador do Distrito do Moxico, "Recrutamento" (July 19, 1932), 1, MAUC, Folder 86 24°.

95. Table of total number of workers contracted by the Agência of Saurimo, pertaining to the Circumscriptions of the District of Lunda (1930), MAUC, Folder 86 22°.

96. Letter from H. J. da Fonseca, "Fuga de trabalhadores" (October 14, 1931), 1, MAUC, Folder 86 23°.

97. Spamoi report (June 1941), 3, MAUC, Folder 86D 2°. Conversely, Ball reports that the desertion rate at the Cassequel Company, on the central Angolan coast, was 9.5 percent in 1965. Jeremy Ball, "'The Colossal Lie': The Sociedade Agrícola do Cassequel and Portuguese Colonial Labor Policy in Angola, 1899–1977" (PhD diss., UCLA, 2003), 219.

98. "Trabalhadores" (February 15, 1930), 1, MAUC, Folder 86 19°.

99. A 1971 company report reveals that Diamang intended to create a centralized, worker database to prevent laborers from employing this tactic. Carreira, Report (July 17, 1971), 4, MAUC, Folder 11-C.

100. The UPA, or União das Populações de Angola, was a Congo-based nationalist group founded in the 1950s. The UPA was eventually subsumed by Holden Roberto's FNLA (Frente Nacional de Libertação de Angola) movement, which survives in contemporary Angola as a marginalized political party.

101. *Relatório imediato* (January 8, 1963), 1, TT, Folder PIDE/DGS Del A—P Inf—10.06.C. Although Cooper persuasively argues that the harsh contract labor system in Namibia ultimately spawned the independence movement SWAPO (South West Africa People's Organization), these types of causal connections are absent in the Diamang context. Allan D. Cooper, "The Institutionalization of Contract Labour in Namibia," *Journal of Southern African Studies* 25, no. 1 (March 1999): 138.

102. Letter from António Baptista Potier to the PIDE (January 16, 1963), 1, TT, Folder PIDE/DGS Del A, P Inf, 11.27.D/1. Specialized workers were Africans who possessed valuable skills, including, for example, heavy equipment operators, mechanics, or clerks. For examples of these types of desertions, see José Lopes, *PIDE Report* (September 12, 1964), TT, Folder PIDE/DGS Del A, P Inf, 11.27.D/1.

103. "Nota de Informação no. 73/72" (August 24, 1972), 1, MAUC, Folder 86A,7 32.

104. Letter from (signature illegible) to the Senhor Ministro do Ultramar (1936): 32, Special Collections Library at the University of Virginia (hereafter UVA).

105. Diamang, *Breve notícia sobre a representação da Companhia de Diamantes de Angola na feira das indústrias portuguesas* (Luanda: Emprésa Gráfica de Angola, 1954), 9.

106. Memorandum on Mine Labor (1938), 7, MAUC, Folder 86 37°.

107. Fernando Duarte, "Inspecção à mão-de-obra" (November 3, 1972), MAUC, Folder 86A, 7 32.

CHAPTER 6: EVENTFUL EVENINGS

1. Diamang created Spamoi (or, after 1962, "Spamo") to assist African employees and their families with all nonworksite facets of life, including health and hygiene, dispute resolution, rations, and housing.

2. Domingos Cazeweque, interview by the author, Lunda (Angola), November 18, 2005.

3. Unfortunately, evidence of strategic pursuits within voluntários' home villages and settlements is too thin and fragmented to include in this examination.

4. By largely withholding accommodations from voluntários, the company prompted these laborers to remain rooted in their proximate home villages and communities. Only from the 1940s did Diamang begin to accommodate small numbers of single or unaccompanied voluntários in its mine encampments.

5. Robert Turrell, *Capital and Labour on the Kimberley Diamond Fields, 1871–1890* (Cambridge: Cambridge University Press, 1987), 155.

6. T. Dunbar Moodie and Vivienne Ndatshe, *Going for Gold: Men, Mines, and Migration* (Berkeley: University of California Press, 1994), 182.

7. In 1936, Diamang formalized a one-week "acclimation period" for workers constructing their own houses. *Relatório—Apresentados pelos administradores por parte do governo na Diamang relativo ao ano de 1933* (1934), 20–21, Arquivo Torre do Tombo (hereafter TT), Folder AOS/CO/UL-8A; Angola, *Código do trabalho dos indígenas* (Luanda: Imprensa Nacional de Angola, 1936), 124.

8. Occasionally, Diamang was forced to reassign workers away from their existing mining jobs in order to assist in the construction of temporary housing, thereby irking mine managers.

9. In reference to this practice, White writes, "As at the Cape, where straight lines of rectangular houses with neat gardens were the sign of Christian dwellings, so at Magomero [Malawi] 'civilization began by squaring the circle.'" Landeg White, *Magomero: Portrait of an African Village* (Cambridge: Cambridge University Press, 1987), 25. See also James C. Scott, *Seeing Like a State: How Certain Schemes to Improve the Human Condition Have Failed* (New Haven, CT: Yale University Press, 1998).

10. As Diamang exhausted individual mines and shifted work sites, this option was not always available.

11. Letter from Jorge Barros to the Gov. Geral de Angola, (March 11, 1943), 7, Museum of Anthropology, Universidade de Coimbra (hereafter MAUC), Folder 86 42°.

12. In some senses, the company's broader "protection zone," which delimited its concessionary area and which it actively monitored, acted as a confinement system and thus provided a type of labor control.

13. Only at its selection stations did Diamang institute anything resembling the compound system, yet this implementation affected only a small number of workers. By the early 1940s, company officials appear to have completely abandoned the notion of implementing a compound(-like) system. In that year, Diamang sent René Delville, a company engineer, to South Africa to study the Rand compounds with instructions to "get an idea of the type of meals supplied to the natives, the manufacture of native beer, and any other thing relative to native labor and welfare," but with no explicit orders

to examine the housing arrangements or configuration. Letter from A. Pinto Ferreira to H. T. Dickinson (July 7, 1942), MAUC, Folder 86 41°.

14. Spamoi report (June 1941), 7, MAUC, Folder 86D 2°, MAUC.

15. Letter from A. Pinto Ferreira to Ernesto de Vilhena (February 11, 1943), 2, MAUC, Folder 86D 3°.

16. Lina Machamba, interview by the author, August 16, 2005, Lunda (Angola). She also indicated that her husband still worked with a shovel, such that after each taxing day he was enjoying "top of the line" employee housing.

17. Romano, *Relatório anual da Spamoi, 1968* (February 20, 1969), 7, MAUC, Folder 86D, 2 20°. Other encampments featured children's parks, pools or even adult recreation centers.

18. João Paulo Sueno, interview by the author, Lunda (Angola), August 12 and 15, 2005. This scenario was common in the western stretches of Lunda in the 1970s, in which new mines were opened hurriedly, in great part to preclude illegal mining operations from removing stones before Diamang had the opportunity to do so.

19. Cafololo Muamuiombo, interview by the author, Lunda (Angola), August 12, 2005.

20. J. K. McNamara, "Brothers and Work Mates: Home Friend Networks in the Social Life of Black Migrant Workers in a Gold Mine Hostel," in *Black Villagers in an Industrial Society*, ed. Philip Mayer (Cape Town: Oxford University Press, 1980), 320.

21. Letter from H. T Dickinson to Colonel de Mello (July 17, 1929), MAUC, Folder 86 18°.

22. Angola, *Código do trabalho dos indígenas* (1936), 15. In this case, the company honored the former stipulation, but occasionally still required children to share a room with their parents.

23. C. Veloso, *SPAMOI relatório anual de 1960* (March 21, 1961), 12, MAUC, Folder 86D, 2 13°; J. Figueiredo, *Relatório anual da Spamoi, 1961* (March 14, 1962), 10, MAUC, Folder 86D, 2 14°.

24. Interviews by the author in Lunda, for example: António Batista, August 9, 2005; Cafololo Gustavo, August 15, 2005.

25. Joaquim António Issuamo, interview by the author, Lunda (Angola), November 19, 2004.

26. Francisco Xamucuco, interview by the author, Lunda (Angola), November 21, 2005.

27. Gordon has examined the tension this mixture generated within Namibian diamond mining communities. Robert J. Gordon, *Mines, Masters and Migrants: Life in a Namibian Compound* (Johannesburg: Ravan Press, 1977), 114. Harries also examines this phenomenon on South African mines, in particular related to new waves of migrant recruits. Patrick Harries, *Work, Culture, and Identity: Migrant Laborers in Mozambique and South Africa, c. 1860–1910* (Portsmouth, NH: Heinemann, 1994), 56.

28. Joaquim António Issuamo, interview by the author, Lunda (Angola), November 19, 2004.

29. Some voluntários constructed temporary housing close to the mines, sleeping in it periodically, though rarely on consecutive nights. Despacho (January 12, 1942), Folder 86D 2°, MAUC.

30. Lute J. Parkinson, *Memoirs of African Mining* (self-published, 1962), 38.

31. Correia Oliveira, "Nota de informação no. 100/70" (May 25, 1970), 3, MAUC, Folder 11-B, 1.

32. Mualesso Gaston, interview by the author, Lunda (Angola), August 9, 2005. Other informants recalled much shorter distances—as little as 200 meters—but most indicated that walks were roughly two to five kilometers.

33. Itela Joaquim, interview by the author, Lunda (Angola), August 12, 2005.

34. No more than 15 percent of workers were ever housed in encampments that featured running water.

35. Sacabela Sacahiavo, interview by the author, Lunda (Angola), August 12, 2005.

36. Martam Camanda, interview by the author, Lunda (Angola), August 16, 2005.

37. Paulo Chingueji, interview by the author, Lunda (Angola), November 22, 2005.

38. Cafololo Muamuiombo, interview by the author, Lunda (Angola), August 12, 2005.

39. Mine bosses were acutely reluctant to allow food shortages to obstruct the absorption of ever-increasing numbers of workers, and thus they prioritized the procurement of food supplies.

40. Diamang, *Extracto de entrevista com o ex-chefe da zôna mineira da unda, engenheiro Sr. Eugenio Salles Lane, publicada no jornal "A Patria"* (Luanda: Empreza Grafíca de Angola, 1925), 12.

41. MAUC, Folder 86 7° (February 1927).

42. Local African women unaffiliated with the company played a central role in this development by plying their produce at weekly, company-organized markets. See *Relatório apresentados pelos administradores* (1934), 11, TT, Folder AOS/CO/UL-8A.

43. A. Pinto Ferreira, "Notas sobre a alimentação dada" (April 13, 1945), 2, MAUC, Folder 86D 4°.

44. Deque, interview by the author, Lunda (Angola), August 10, 2005.

45. Barun Mitra, "Don't copy western models of conservation. The lord of the jungle can be saved if commerce is harnessed to his cause." *Indian Express* (Delhi edition), March 18, 2005.

46. One possibility for the archival silence prior to 1962 is that the size or amount of rations distributed by the company was at least equal, if not superior, to what African laborers were used to consuming.

47. Figueiredo, *Relatório anual da Spamoi, 1962* (March 4, 1963), 21, MAUC, Folder 86D, 2 15°.

48. Ibid. *Bombó* is manioc after it has been treated in water but before it is ground into fuba.

49. Marvanejo (February 6, 1963), 4, MAUC, Folder 86D 6°.

50. I would suggest that this activity transpired largely undetected and, thus, would not have featured in company records or perhaps was simply tolerated by sentinelas and thus went unreported.

51. Filipe Saucauenhe, interview by the author, Lunda (Angola), November 18, 2005.

52. João Paulo Sueno, interview by the author, Lunda (Angola), August 12 and 15, 2005.

53. Parkinson, *Memoirs of African Mining*, 43–44.

54. Dinis dos Santos Muriandambo, interview by the author, Lunda (Angola), August 11, 2005.

55. Penvenne has astutely argued that the materialist bent of labor history has undervalued people's attention to their social and spiritual needs. Jeanne Marie Penvenne, *African Workers and Colonial Racism: Mozambican Strategies and Struggles in Lourenço Marques, 1877–1962* (Portsmouth, NH: Heinemann, 1995), 10.

56. Dinis dos Santos Muriandambo, interview by the author, Lunda (Angola), August 11, 2005.

57. Charles van Onselen, *Chibaro: African Mine Labour in Southern Rhodesia, 1900–1933* (London: Pluto Press, 1976), 177.

58. Dinis dos Santos Muriandambo, interview by the author, Lunda (Angola), August 11, 2005. Although Muriandambo's assessment appears to be exaggerated, the fact that single men pursued married women is incontrovertible.

59. Paulino, interview by the author, Lunda (Angola), August 11 and 12, 2005.

60. Paulo Chingueji, interview by the author, Lunda (Angola), November 22, 2005.

61. Txipanda Armando, interview by the author, Lunda (Angola), November 21, 2005.

62. José Silva, interview by the author, Lunda (Angola), August 12, 2005.

63. Caiombo Jombe, interview by the author, Lunda (Angola), August 9, 2005.

64. Mateus Nanto, interview by the author, Lunda (Angola), August 12, 2005.

65. Martam Camanda, interview by the author, Lunda (Angola), August 16, 2005.

66. Bernardo Montaubuleno, interview by the author, Lunda (Angola), November 20, 2004.

67. Paulino, interview, Lunda (Angola), August 11 and 12, 2005.

68. Dinis dos Santos Muriandambo, interview by the author, Lunda (Angola), August 11, 2005.

69. António Muiege, interview by the author, Lunda (Angola), November 21, 2005.

70. Heidi Gengenbach's examination of the ways that African women in colonial Mozambique forged chains of feminine fellowship that transcended

boundaries of marriage, class, ethnicity, race, and nationality informs my examination of women's strategic pursuit of camaraderie at Diamang. Heidi Gengenbach, "Boundaries of Beauty: Tattooed Secrets of Women's History in Magude District, Southern Mozambique," *Journal of Women's History* 14, no. 4 (2003): 135.

71. Mulevana Camachele, interview by the author, Lunda (Angola), August 16, 2005.

72. Mawassa Mwaninga, interview by the author, Lunda (Angola), November 22, 2005.

73. Felipe Leo Muatxissupa, interview by the author, Lunda (Angola), August 15, 2005.

74. Itela Joaquim, interview by the author, Lunda (Angola), August 12, 2005.

75. Janette Pedro, interview by the author, Lunda (Angola), August 16, 2005. In at least one instance, a mine encampment helped unite a couple in marriage. In this scenario, Aida Fernando met her future husband there. See epigraph for further detail. Aida Fernando, interview by the author, Lunda (Angola), November 22, 2005.

76. João Paulo Sueno, interview by the author, Lunda (Angola), August 12 and 15, 2005.

77. João Saluembe, interview by the author, Lunda (Angola), November 21, 2005.

78. Bernardo Montaubuleno, interview by the author, Lunda (Angola), November 20, 2004.

79. The copious photos of kids playing on company-installed recreational equipment in the Diamang archive are somewhat misleading, however, as only newly constructed settlements featured this type of equipment.

80. Deque, interview by the author, Lunda (Angola), August 10, 2005.

81. Mulevana Camachele, interview by the author, Lunda (Angola), August 16, 2005.

82. João Paulo Sueno, interview by the author, Lunda (Angola), August 12 and 15, 2005.

83. Caiombo Jombe, interview by the author, Lunda (Angola), August 9, 2005.

84. Deque, interview by the author, Lunda (Angola), August 10, 2005.

85. Paulino, interview by the author, Lunda (Angola), August 11 and 12, 2005.

86. Filipe Saucauenhe, interview by the author, Lunda (Angola), November 18, 2005.

87. African employees were also welcome to visit Diamang's museum in Dundo, launched in 1936, which featured "native folklore and artifacts." However, as the museum was located in Dundo and those in transit, either coming or going, were not permitted to visit, it was frequented by very few mine workers and thus remains outside the focus of this study. See Nuno Porto, *Angola a preto e branco: Fotografia e ciência no museu do Dondo, 1940–1970* (Coimbra: Museu Antropológico da Universidade de Coimbra,

1999); Nuno Porto, "Manageable Past: Time and Native Culture at the Dundo Museum in Colonial Angola," *Cahiers d'Études Africaines* 39, no. 155 (1999): 767–87.

88. Porto, "Manageable Past," 778.

89. Beginning with the 1952 Festa, the company publicly feted veteran employees with gold, silver, and copper medals (according to shifting longevity thresholds), as well as with small financial bonuses. By 1963, Diamang had distributed over 6,000 medals (510 gold, 1,647 silver, and 3,779 copper) and over 3,760$00 escudos in bonuses. Diamang, *Argumento do filme intitulado "O romance do luachimo—Lunda, terra de diamantes"* (Lisbon: Companhia de Diamantes, 1963).

90. J. Robalo, *Relatório mensal de Julho de 1951* (August 14, 1951), 10, Folder 86D, 2 5°, MAUC.

91. Fernando Meuaçefo, interview by the author, Lunda (Angola), August 11, 2005.

92. Diamang, *Breve notícia, sobre a sua actividade em Angola* (Lisbon: Tip. Silvas, 1963), 55.

93. SPAMOI, *Relatório anual de 1953* (January 27, 1954), MAUC, Folder 86D, 2 7°. Screenings of these films allegedly prompted workers to articulate both their desire to travel to Portugal and their understanding of why Portuguese workers would want to return there.

94. MAUC, Folder 86D, 2 9° (1956).

95. SPAMOI, *Relatório anual de 1953* (January 27, 1954), MAUC, Folder 86D, 2 7°. Films were similarly popular on South Africa's mines, though were managed by missions instead of mining companies. Tim Couzens, "'Moralizing Leisure Time': The Transatlantic Connection and Black Johannesburg, 1918–1936," in *Industrialisation and Social Change in South Africa: African Class Formation, Culture, and Consciousness, 1870–1930*, ed. Shula Marks and Richard Rathbone (New York: Longman, 1982), 321.

96. Only when Diamang screened the same films repeatedly or when they were in poor condition or offered deficient translations did attendance drop, though never dramatically.

97. MAUC, Folder 86 48° (1946).

98. Reunião de chefes de grupo (April 9, 1970), 4, MAUC, Folder 86B, 6°.

99. Emmanuel Akyeampong, *Drink, Power, and Cultural Change: A Social History of Alcohol in Ghana, c. 1800 to Recent Times* (Portsmouth, NH: Heinemann, 1996), 4.

100. João Paulo Sueno, interview by the author, Lunda (Angola), August 12 and 15, 2005.

101. As with most scenarios and settings beyond the company's gaze, Diamang officials typically only mentioned these celebratory activities when problems, such as increased absences on Monday, resulted.

102. Parkinson, *Memoirs of African Mining*, 52.

103. Bernardo Montaubuleno, interview by the author, Lunda (Angola), November 20, 2004.

104. Isabela Casombe, interview by the author, Lunda (Angola), November 22, 2005.

105. Luciano Xacambala, interview by the author, Lunda (Angola), November 21, 2005. As Pongweni has demonstrated in his study of songs employed during the Zimbabwean war of liberation, songs can serve as a binding force when they emphasize common, challenging conditions. Alec J. C. Pongweni, "The Chimurenga Songs of the Zimbabwean War of Liberation," in *Readings in African Popular Culture*, ed. Karin Barber (Bloomington: University of Indiana Press, 1997), 67.

106. For similar examples from colonial Mozambique, see Leroy Vail and Landeg White, "Plantation Protest: The History of a Mozambican Song," in *Readings in African Popular Culture*, ed. Karin Barber (Bloomington: University of Indiana Press, 1997), 63.

107. Miudo Rafael, interview by the author, Lunda (Angola), August 15, 2005.

108. António Batista, interview by the author, Lunda (Angola), August 9, 2005. See chapter 2 for further information about the arrival of PIDE in Lunda.

109. Between 1911 and 1922, Norton de Matos, governor-general and then high commissioner, had promulgated a range of colonial legislation restricting Africans from purchasing or manufacturing alcohol.

110. Caiombo Jombe, interview by the author, Lunda (Angola), August 9, 2005.

111. João Paulo Sueno, interview by the author, Lunda (Angola), August 12 and 15, 2005. This willingness to part with a portion of their rations suggests that the provision was more than adequate. In a case from 1963, for example, an individual named José Mualege was apprehended for participating in a clandestine market and allegedly explained: "We were not lacking for fish, but I wanted tobacco because the company did not provide me with any." Romano and Mendes, *Spamoi Report* (March 3, 1963), 5, MAUC, Folder 86D 6°.

112. Joaquim António Issuamo, interview by the author, Lunda (Angola), November 19, 2004.

113. Mawassa Mwaninga, interview by the author, Lunda (Angola), November 22, 2005.

114. In fact, sentinelas often enjoyed close relationships with encampment residents, socializing with them at night, and especially on Saturday evenings. Although the power relationship in these instances was clearly asymmetrical, both company records and informants' testimony suggest that arbitrary violence was rare.

115. Joaquim António Issuamo, interview by the author, Lunda (Angola), November 19, 2004.

116. Romano, *Relatório anual da SPAMO, 1967* (February 10, 1968), 21, Folder 86D, 2 19°, MAUC.

117. Joaquim António Issuamo, interview by the author, Lunda (Angola), November 19, 2004.

118. Lina Machamba, interview by the author, Lunda (Angola), August 16, 2005.

119. Companhia de Diamantes de Angola, Relatório da Viagem a Africa do Administrador Dr. J. Simões Neves em 1945, TT, Folder AOS/CO/UL-8A1.

120. Parkinson, *Memoirs of African Mining*, 43–44.

121. Dinis dos Santos Muriandambo, interview by the author, Lunda (Angola), August 11, 2005.

122. Pragmatic reasons also played a role: sobas helped furnish laborers for Diamang's mines, whereas company officials regarded "medicine men" as potential troublemakers.

123. Dinis dos Santos Muriandambo, interview by the author, Lunda (Angola), August 11, 2005. Muriandambo added that "there were times, however, when secrets of a certain ceremony were missing because no one there knew them, but it would continue because of a deep cultural knowledge, feeling, and understanding, and [because of] the spirit of the group that was present."

124. For a sustained discussion of the role of radio in the development of Angolan nationalism, see Marissa Moorman, *Intonations: A Social History of Music and Nation in Luanda, Angola, from 1945 to Recent Times* (Athens: Ohio University Press, 2008).

125. Albuquerque Matos, "Comunicações," in Reis, *Diamante* (2003): 15.

126. For those who did listen, the two most popular transmissions were the MPLA's "Angola Combatante" and the FNLA's "Angola Livre." Broadcasts were aired in French, Portuguese, and a variety of local languages.

127. Joaquim Trinidade, interview by the author, Lunda (Angola), November 15, 2004.

128. Silvestre Muachembe, interview by the author, Lunda (Angola), November 19, 2004.

129. The MPLA is an initialism for the Movimento Popular de Libertação de Angola, or Popular Movement for the Liberation of Angola.

130. Bartolomeu Lubano, interview by the author, Lunda (Angola), August 10, 2005. Two anecdotes are testament to the ferocity of the PIDE. Portuguese employee Joaquim Macedo recalled: "Our house [in Dundo] was near the PIDE [office]. We often heard screams at night. The kids said that PIDE agents used to take bodies of dead prisoners at night to throw into the Luachimo [river]. . . . We used to pass there scared and at a prudent distance." Joaquim Macedo, "Dundinices," in Bernardo Reis and Vítor Sousa, eds., *Diamante*, XVII encontro Diamang (Braga, 1999): 24. Meanwhile, Fernando Simão, an Angolan who also lived around Dundo, declared that "PIDE would regularly beat people up, kill people; bodies would be found in the river. You could hear the torture from the neighborhoods around Dundo." Fernando Simão, interview by the author, Lunda (Angola), August 10, 2005.

1. Letter from G. H. Newport to the Bureau Technique (June 18, 1925), 2, Museum of Anthropology, Universidade de Coimbra (hereafter MAUC), Folder 86 3°.

2. Unfortunately, Diamang did not keep statistics regarding the numbers or percentages of married couples who opted to stay on, even though it hoped for exactly this scenario, and evidence from mining companies on the Copperbelt suggests that this practice was (more) common. See Charles Perrings, *Black Mineworkers in Central Africa: Industrial Strategies and the Evolution of an African Proletariat in the Copperbelt, 1911–41* (New York: Africana, 1979), 82; Jane L. Parpart, *Labor and Capital on the African Copperbelt* (Philadelphia: Temple University Press, 1983), 35, 42.

3. *Spamoi report* (June 1941), 1, MAUC, Folder 86D 2°. These incentives resembled similar retention enticements tendered over time by mining companies on the Copperbelt.

4. *Spamoi report* (June 1941), 2, MAUC, Folder 86D 2°.

5. Joaquim António Issuamo, interview by the author, Lunda (Angola), November 19, 2004.

6. Cafololo Gustavo, interview by the author, Lunda (Angola), August 15, 2005. Gustavo explained that a friend of his with whom he had traveled to Diamang also decided to stay, but that six others elected to leave.

7. A. C. Figueira, *Spamoi 1937 relatório annual* (February 9, 1938), 1, MAUC, Folder 86D, 2 1°.

8. Figueira, *Spamoi 1937 relatório annual* 6, MAUC, Folder 86D, 2 1°.

9. Manuel Pereira Figueira, *Relato sôbre mão d'obra indígena* (May 18, 1938), 39, MAUC, Folder 86 36°. In a similar incident in 1937, fifty-four workers had opted to remain on, but when they asked for their families to join them, only two of their wives allegedly declared their willingness to do so to local authorities.

10. Letter from J. Tavares Paulo to Ernesto de Vilhena (April 28, 1950), MAUC, Folder 86B, 6 3°.

11. Letter from J. Tavares Paulo to the Agente no Chitato (August 4, 1947), 1–2, MAUC, Folder 86 51°.

12. Joaquim António Issuamo, interview by the author, Lunda (Angola), November 19, 2004.

13. Fernando Tximvula, interview by the author, Lunda (Angola), August 9, 2005.

14. Letter from G. H. Newport to Major Torre do Valle (August 24, 1925), 1, MAUC, Folder 86 3°.

15. Diamang, *Relatório da viagem a Africa do administrador Dr. J. Simões Neves* (1945), 107, Arquivo Torre do Tombo (hereafter TT), Folder AOS/CO/UL-8A1. It is unclear how successful these early bids to have voluntários renew oral contracts or simply keep showing up for work each Monday were, but they do highlight the company's efforts and officials' underlying beliefs that workers would respond if only offered the right mix of incentives.

16. J. Janmart, *Relatório geral de prospecção, Julho de 1948* (August 4, 1948), MAUC, Folder 86B,6 1ª.

17. Noronha Feyo, "Nota de informação no. 28/46" (October 9, 1946), MAUC, Folder 86D 4°.

18. In the early 1920s, the company was withholding and sending on to postos for collection the entirety of workers' wages, while by the mid-1920s this figure was reduced to 80 percent, and thereafter to 50 percent, before being set at 60 percent for the remainder of the contratado period.

19. Ross chronicled instances of this scheme during his 1925 investigatory trip to Angola, while Heywood has examined this phenomenon in Angola's central highlands. Edward Alsworth Ross, *Report on Employment of Native Labor in Portuguese Africa* (New York: Abbott Press, 1925), 9; Linda Heywood, *Contested Power in Angola, 1840s to the Present* (Rochester, NY: University of Rochester Press, 2000), 50.

20. MAUC, Folder 86 7° (1927). The state also absurdly alleged that workers would "be unaware of, and thus potentially unable to meet, any new taxes that had been imposed while they were away at the mines."

21. Letter from M. K. Shaler to Brandão de Mello, "Travailleurs" (June 13, 1921), MAUC, Folder 86 1°.

22. Letter from Brandão de Mello to Ernesto de Vilhena, "Queixa do administrador de lungué bungo" (September 9, 1930), MAUC, Folder 86 21° (1930).

23. Diamang Serviço de Saúde, *Relatório sôbre inspecção da indígenas no distrito da Lunda, inspecções da indígenas nas circunsrições e postos* (1938), 13, MAUC, Folder 86 38°.

24. Luciane Kahanga, interview by the author, Lunda (Angola), November 18, 2005.

25. In an incident from 1947, sobas from the Saurimo area accused a chefe de posto named Angleu Teixeira of pocketing salaries owed to workers. In turn, Diamang officials requested that he be transferred (Letter from Silvio Guimarães to Ernesto de Vilhena [August 21, 1947], MAUC, Folder 86 51°). In 1971, an ILO investigation uncovered an episode in which a chefe had unsuccessfully requested that a returning worker pay his tax after he had already paid it at Diamang. ILO, *Report on Direct Contacts with the Government of Portugal regarding the Implementation of the Abolition of Forced Labour Convention, 1957* (1971), 25.

26. This particular official was actively campaigning for the company to reduce contracts from eighteen to twelve months. Letter from an unidentified colonial official to the Overseas Minister (1952), 26, Special Collections Library at the University of Virginia (hereafter UVA).

27. Mateus Nanto, interview by the author, Lunda (Angola), August 12, 2005.

28. Cafololo Muamuiombo, interview by the author, Lunda (Angola), August 12, 2005.

29. Muhetxo Sapelende, interview by the author, Lunda (Angola), August 11, 2005.

30. Domingos Cazeweque, interview by the author, Lunda (Angola), November 18, 2005.

31. Luciane Kahanga, interview by the author, Lunda (Angola), November 18, 2005.

32. António Muiege, interview by the author, Lunda (Angola), November 21, 2005.

33. Filipe Saucauenhe, interview by the author, Lunda (Angola), November 18, 2005.

34. Mateus Nanto, interview by the author, Lunda (Angola), August 12, 2005.

35. Manuel Pereira Figueira, *Relato sôbre mão d'obra indígena* (May 18, 1938), 4, MAUC, Folder 86 36°; Diamang, *Relatório de conselho de administração* (Luanda: Diamang, 1953), 38. Emphasis mine.

36. *Informação para o Sr. Director Geral* (October 1962), 7, MAUC, Folder 86A,2a 2°.

37. Costa Chicungo, interview by the author, Lunda (Angola), May 13, 2005.

38. Legislation introduced in 1925 and reaffirmed in 1938 stipulated that workers were entitled to a rest period equivalent to at least the amount of time they had just served—legislation that state recruiters often flouted. Antero Carvalho, *Instruções provisórias para a recrutamento e emprêgo na província* (Luanda: I. Nacional, 1925), 17; Letter from Raúl Mesquita to Diamang (July 14, 1938), 1, MAUC—Folder 86 36°.

39. ILO Official Bulletin, # 45 (XLV) 1962. Supplement II, No. 2 (April 1962), 11.

40. Muhetxo Sapelende, interview by the author, Lunda (Angola), August 11, 2005.

41. João Muacasso, interview by the author, Lunda (Angola), August 11, 2005.

42. Itela Joaquim, interview by the author, Lunda (Angola), August 12, 2005.

43. Rodrigues, interview by the author, Lunda (Angola), August 16, 2005. See the epigraph of this chapter.

44. Mulevana Camachele, interview by the author, Lunda (Angola), August 16, 2005.

45. Augusto Funete, interview by the author, Lunda (Angola), November 21, 2005.

46. Letter from J. F. de Barros to E. S. Lane, "Queixas" (July 5, 1928), MAUC, Folder 86 12 °.

47. Letter from José Pires de Mendonça to the Snr. Agente do Curador no Chitato (August 20, 1951), 2, MAUC, Folder 86 56°. Competition from sugar companies in this area dates to the 1920s.

48. Letter from an unidentified colonial official to the Senhor Ministro do Ultramar (1952), UVA.

49. Letter from Carlos Borges de Sousa to J. Tavares Paulo (November 7, 1947), MAUC, Folder 86 51°.

EPILOGUE

1. Fighting between the independence movements had already begun prior to 1975. See William Minter, ed., *Operation Timber: Pages from the Savimbi Dossier* (Trenton, NJ: Africa World Press, 1988).

2. Daniel Catuesse, interview by the author, Lunda (Angola), November 21, 2005.

3. Isabel Reis, interview by the author, Oeiras (Portugal), June 15, 2004.

4. For further information, see Shawn McCormick, *The Angolan Economy: Prospects for Growth in a Postwar Environment* (Washington, DC: Center for Strategic and International Studies, 1994), 29; Filip de Boeck, "Borderland Breccia: The Mutant Hero in the Historical Imagination of a Central-African Diamond Frontier," *Journal of Colonialism and Colonial History* 1, no. 2 (2000): 15.

5. Bernardo Reis, "O ultimo arrear da Bandeira Portuguesa no Dundo," in *Diamante* XXI encontro Diamang (2003): 80.

6. During this transitional period, Dundo became a frequent destination for elite members of the MPLA, including Agostinho Neto, who would become Angola's first president after his party seized Luanda and declared itself the country's sole legitimate government.

7. By Decree No. 61/77 of August 24, 1977, Angola nationalized the stock of all small shareholders, as Angola's president, Agostinho Neto, accused them of ownership for the sole purpose of collecting dividends. This move raised the Angolan state's ownership of Diamang's joint stock to 68.85 percent.

8. Diamang, *Relatório e projecto de cisão dos bens situados em Portugal que pertenceram à Companhia de Diamantes de Angola, S. A. R. L.* (Lisbon: C. D. A., 1979): 7. Oddly, Neto's math indicates that he considered Diamang to have commenced operations only in 1921, the year of the landmark, far-reaching state-company agreement, rather than in 1917.

9. Rafael Marques, an Angolan journalist, has chronicled instances of this violence, though many more episodes have gone unreported. See Rafael Marques, *Angola's Deadly Diamonds: Lundas, the Stones of Death* (2005); Rafael Marques, *Operação kissonde: Os diamantes da humilhação e da miséria* (2006); Rafael Marques, *Diamantes de sangue: Corrupção e tortura em Angola* (Lisbon: Tinta da China, 2011).

10. Isabel Reis, interview by the author, Oeiras (Portugal), June 15, 2004.

Bibliography

INTERVIEWS

ANGOLA

Aida Fernando, November 22, 2005
Alberto Rossa, November 16, 2004
Andre Muamukepe, May 18, 2005
Anna Maria Dos Santos, August 10, 2005
António Batista, August 9, 2005
António Cuassue, November 19, 2004
António Muiege, November 21, 2005
António Sulessa, May 17, 2005
Araújo Caxide, August 15, 2005
Armando Nguatje, November 19, 2004
Augusto Funete, November 21, 2005
Augusto Kalanje, November 17, 2004
Bartolomeu Lubano, August 10, 2005
Bernardo Montaubuleno, November 20, 2004, and May 18, 2005
Cafololo Gustavo, August 15, 2005
Cafololo Muamuiombo, August 12, 2005
Caiombo Jombe, August 9, 2005
César Francisco Adão Categória, August 9, 2005
Costa Chicungo, May 13, 2005
Daniel Catuesse, November 21, 2005
David Muangongo, November 22, 2005
Deque, August 10, 2005
Dinis dos Santos Muriandambo, August 11, 2005
Domingos Cazeweque, November 18, 2005
Domingos Matos, November 18, 2005
Eugenia Monteiero, October 14, 2004
Feijão, November 20, 2005
Felipe Leo Muatxissupa, August 15, 2005
Fernando Meuaçefo, August 11, 2005
Fernando Simão, August 10, 2005
Fernando Tximvula, August 9, 2005.

Filipe Saucauenhe, November 18, 2005
Fina da Costa, May 7, 2005
Francisco Xamucuco, November 21, 2005
Gabriel Alberto, August 16, 2005
Francisco Gomes Maiato, May 13, 2005
Isabela Casombe, November 22, 2005
Itela Joaquim, August 12, 2005
Janette Pedro, August 16, 2005
João Kakesse, November 19, 2004
João Muacasso, August 11, 2005
João Paulo Sueno, August 12 and 15, 2005
João Saluembe, November 21, 2005
João Zecai, August 9, 2005
Joaquim António Issuamo, November 19, 2004
Joaquim Ezaia, August 10 and 11, 2005
Joaquim Muamungo, November 18, 2005.
Joaquim Trinidade, November 15, 2004
José Coxi, August 16, 2005
José Silva, August 12, 2005
José Turiambe Muachiriango, August 15, 2005
Lina Machamba, August 16, 2005
Lucas Macafuela, August 16, 2005
Luciane Kahanga, November 18, 2005
Luciano Xacambala, November 21, 2005
Major Agostinho, November 21, 2005
Mário Alfredo Samuhaniquime, November 18, 2005
Martam Camanda, August 16, 2005
Mateus Nanto, August 12, 2005
Mawassa Mwaninga, November 22, 2005
Miudo Rafael, August 15, 2005
Mualesso Gaston, August 9, 2005
Muatxinjango Maca, November 21, 2005
Muatxissengue, November 22, 2005
Muhetxo Sapelende, August 11, 2005
Mulevana Camachele, August 16, 2005
Mulombe Manuel, August 10, 2005.
Paulino, August 11 and 12, 2005
Paulo Caninda, November 19, 2004
Paulo Chingueji, November 22, 2005
Paulo Leão Vega, May 17, 2005
Pedro Bento Marques, August 10, 2005
Rodrigo Lino, May 17, 2005
Rodrigues, August 16, 2005
Sacabela Sacahiavo, August 12, 2005
Silvério Monteiro, July 27, 2005

Silvestre Muachembe, November 19, 2004
Teresa Penedo, November 16, 2004
Txipanda Armando, November 21, 2005
Vitória Trinidade, November 15, 2004

PORTUGAL

Carlos Machado, March 18, 2006
Isabel Reis, June 15, 2004
José Gradil, June 12, 2004
Julietta Machado, March 18, 2006
Leonardo Manuel Judas Chagas, June 7, 2006

PERIODICALS AND NEWSPAPERS

A Província de Angola
Boletim Geral das Colónias. nos. 152 and 162 (1938), 242–43 (1945), 305 (1950)
Boletim Oficial da Província de Angola, series 1, no. 20 (May 14, 1921): 37
O Comércio (Porto, Portugal)
O Comércio: Jornal de Propaganda e Defesa da Actividade Econômica de Angola
Diário de Luanda
Diário de Notícias
Indian Express (Delhi Edition), March 18, 2005
Jornal de Benguela
Jornal do Comércio
O Primeiro de Janeiro
O Século

ARCHIVES

LISBON, PORTUGAL

Arquivo Histórico Ultramarino (AHU)
Sociedade de Geografia de Lisboa (SGL)
Torre do Tombo (TT)

COIMBRA, PORTUGAL

Museum of Anthropology, Universidade de Coimbra (MAUC)

LUANDA, ANGOLA

Arquivo Histórico Nacional (AHN)

CHARLOTTESVILLE, VIRGINIA

University of Virginia Special Collections Library (UVA)

SECONDARY WORKS

Abrantes, Tito Castello Branco. *Diamantes de Angola para Angola*. Lisbon: Neogravura, Lda, 1973.

Abshire, David M., and Michael A. Samuels, eds. *Portuguese Africa: A Handbook*. New York: Praeger, 1969.

Akyeampong, Emmanuel. *Drink, Power, and Cultural Change: A Social History of Alcohol in Ghana, c. 1800 to Recent Times*. Portsmouth, NH: Heinemann, 1996.

Andrade, Carlos Freire de. *Diamond Deposits in Lunda: A Geological Survey Made in 1945–46, Part I*. Lisbon: Diamang, 1953.

Atkins, Keletso. *The Moon Is Dead! Give Us Our Money!: The Cultural Origins of an African Work Ethic, Natal, South Africa, 1843–1900*. Portsmouth, NH: Heinemann, 1993.

Austen, Ralph A., and Rita Headrick. "Equatorial Africa under Colonial Rule." In *History of Central Africa*, vol. 2, edited by David Birmingham and Phyllis M. Martin, 27–94. London: Longman, 1983.

Ball, Jeremy. *Colonial Labor in Twentieth-Century Angola*. www.history compass.com: 2005.

——. "'The Colossal Lie': The Sociedade Agrícola do Cassequel and Portuguese Colonial Labor Policy in Angola, 1899–1977." PhD diss., UCLA, 2003.

——. "'A Time of Clothes': The Angolan Rubber Boom, 1886–1902." *Ufahamu* 28, no. 1 (2000): 25–42.

Beetz, P. F. W. *Preliminary and Final Report on the Angola and Belgian Congo Diamond Fields*. 1929.

Behar, Ruth. *The Vulnerable Observer: Anthropology That Breaks Your Heart*. Boston: Beacon Press, 1996.

Boavida, Américo. *Angola—cinco séculos de exploração portuguêsa*. Rio de Janeiro: Editôra Civilização Brasileira, 1967.

Botelho, António. *Diamang, guia do empregado da mina*. Diamang, 1971.

Burawoy, Michael. *The Politics of Production: Factory Regimes under Capitalism and Socialism*. London: Verso, 1985.

Cadbury, William A. *Labour in Portuguese West Africa*. London: George Routledge and Sons, 1910.

Cameron, Elisabeth L. "Potential and Fulfilled Woman: Initiations, Sculpture, and Masquerades in Kabompo District, Zambia." In *Chokwe!: Art and Initiation Among Chokwe and Related Peoples*, edited by Manuel Jordán, 77–84. New York: Prestel, 1998.

Campos, Eng. Bernardo. "Os kimberlitos e os diamantes: Génese e caracterização." *"Endiama Hoje" Revista Informativa de Empresa Nacional de Diamantes de Angola* (July–August 2004): 18–22.

Carney, Judith A. *Black Rice: The African Origins of Rice Cultivation in the Americas*. Cambridge, MA: Harvard University Press, 2001.

Carreira, António. *Angola: Da escravatura ao trabalho livre: Subsídos para a história demográfica do século XVI até à independência*. Lisbon: Editoria Arcádia, 1977.

Carruthers, Jane. "Creating a National Park, 1910 to 1926." *Journal of Southern African Studies* 15, no. 2 (1989): 188–216. http://dx.doi.org/10.1080/03057078908708197.

Carstens, Peter. *In the Company of Diamonds: De Beers, Kleinzee, and the Control of a Town.* Athens: Ohio University Press, 2001.

Carvalho, Antero Tavares de. *Instruções provisórias para a recrutamento e emprêgo de trabalhadores indígenas na província: Aprovadas por portaria provincial no. 4, de 16 de Janeiro de 1925.* Luanda: Imprensa Nacional, 1925.

Chauncey, George, Jr. "The Locus of Reproduction: Women's Labour in the Zambian Copperbelt, 1927–1953." *Journal of Southern African Studies* 7, no. 2 (1981): 135–64. http://dx.doi.org/10.1080/03057078108708024.

Clarence-Smith, W. G. "Capital Accumulation and Class Formation in Angola." In *History of Central Africa*, vol. 2. edited by David Birmingham and Phyllis M. Martin, 163–99. London: Longman, 1983.

Cleveland, Todd. "Appraising the Value of History: Fieldwork Strategies, Solutions and Lessons from Angola's Diamondiferous Lunda Region, 2004–6." In *Immigrant Academics and Cultural Challenges in a Global Environment*, edited by Femi Kolapo, 239–62. Amherst, NY: Cambria Press, 2008.

———. "The Life of a Portuguese Colonialist: General José Norton de Matos (1867–1955)." MA thesis, University of New Hampshire, 2000.

———. "A Minority in the Middle: Ethnic Baluba, the Portuguese Colonial State, and the Companhia de Diamantes de Angola (Diamang)." In *Minorities and the State in Africa*, edited by Michael Ubanaso and Chima Korieh, 195–215. Amherst, NY: Cambria Press, 2010.

———. "Minors in Name Only: Child Laborers on the Diamond Mines of the *Companhia de Diamantes de Angola* (Diamang), 1917-1975." *Journal of Family History* 35, no. 1 (January 2010): 91–110. http://dx.doi.org/10.1177/0363199009348373.

———. "Rock Solid: African Laborers on the Diamond Mines of the Companhia de Diamantes de Angola (Diamang), 1917–1975." PhD diss., University of Minnesota, 2008.

Coelho, Mário. *"Na terra dos diamantes: Novela Africana."* Lisboa: *Editorial Minerva*, n.d.

Cohen, Robin. "Resistance and Hidden Forms of Consciousness Amongst African Workers." *Review of African Political Economy* 7, no. 19 (September–December 1980): 8–22. http://dx.doi.org/10.1080/03056248008703437.

Cooper, Allan D. "The Institutionalization of Contract Labour in Namibia." *Journal of Southern African Studies* 25, no. 1 (March 1999): 121–38. http://dx.doi.org/10.1080/030570799108786.

Cooper, Frederick. *Decolonization and African Society: The Labor Question in French and British Africa.* Cambridge: Cambridge University Press, 1996. http://dx.doi.org/10.1017/CBO9780511584091.

Costa, Cândido Ferreira da. *Cem anos dos missionários do Espírito Santo em Angola (1866–1966).* Angola: Nova Lisboa, 1970.

Costa, Justo da. *Diamantes de Angola: Nova história de David e Golias.* Braga: Livraria Editora Pax, 1973.

Couzens, Tim. "'Moralizing Leisure Time': The Transatlantic Connection and Black Johannesburg, 1918–1936." In *Industrialisation and Social Change in South Africa: African Class Formation, Culture, and Consciousness, 1870–1930,* edited by Shula Marks and Richard Rathbone, 314–37. New York: Longman, 1982.

Crush, Jonathan, Alan Jeeves, and David Yudelman. *South Africa's Labor Empire: A History of Black Migrancy to the Gold Mines.* Boulder, CO: Westview Press, 1991.

David, J. H. Santos. *Companhia de Diamantes de Angola, direcção dos serviços de Saúde: A endemia tuberculosa na Lunda e Songo.* Dundo: Diamang, 1956.

_____. *Companhia de Diamantes de Angola, direcção dos serviços de Saúde, "Recenseamento demográfico e prospecção sanitária do nordeste da Lunda—1972".* Dundo: Diamang, 1973.

Davidson, Basil. *The African Awakening.* Oxford: Alden Press, 1955.

_____. *In the Eye of the Storm: Angola's People.* New York: Longman, 1972.

Davis, Jennifer, George M. Houser, Susan Rogers, et al. *No One Can Stop the Rain: Angola and the MPLA.* New York: Africa Fund, 1976.

de Boeck, Filip. "Borderland Breccia: The Mutant Hero in the Historical Imagination of a Central-African Diamond Frontier." *Journal of Colonialism and Colonial History* 1, no. 2 (2000): 1–44. http://dx.doi.org/10.1353/cch.2000.0010.

_____. "Domesticating Diamonds and Dollars: Identity, Expenditure and Sharing in Southwestern Zaire (1984–1997)." *Development and Change* 29, no. 4 (1998): 777–810. http://dx.doi.org/10.1111/1467-7660.00099.

de Mello, Antonio Brandão. *Os Diamantes de Angola: Artigos publicados no jornal "A Provincia de Angola" em Março e Abril de 1925.* Luanda: Diamang, 1925.

de Vilhena, Ernesto. *Aventura e rotina: Crítica de uma crítical.* Luanda: Emprêsa Gráfica de Angola, 1954.

Diamang. *Acordo colectivo de trabalho: Do pessoal ao serviço da Companhia de Diamantes de Angola, S.A.R.L.* Lisbon: Escritorio, 1975.

_____. *Actas da reunião da assembleia geral (ordinária) dos accionistas da Companhia de Diamantes de Angola.* Lisbon: 1957, 1959, 1960, 1963–65.

_____. *Algumas notas sôbre a sua actividade.* Dundo: Diamang, 1945.

_____. *Angola: Breve monografia histórica, geográfica e económica, elaborada para a exposição Portuguêsa em Sevilha* (Seville Exposición Ibero-americana) . Loanda: 1929.

_____. *Argumento do filme intitulado "O romance do luachimo—Lunda, terra de diamantes."* Lisbon: Companhia de Diamantes, 1963.

_____. *Breve notícia sobre a representação da Companhia de Diamantes de Angola na feira das indústrias Portuguesas.* Luanda: Emprésa Gráfica de Angola, 1954.

_____. *Breve notícia sobre a sua actividade em Angola.* Lisbon: Tip. Silvas, 1963.

_____. *Condição dos trabalhadores da Lunda.* Luanda: Diamang, 1974.

_____. *Discursos proferidos pelo Sr. Administrador-Delegado, em 21 de Setembro de 1969, na Lunda, por ocasião das solenidades em memória do Comandante Ernesto de Vilhena, Coronel Brandão de Melo e Engenheiro Quirino da Fonseca, bem como algumas palavras dirigidas a sua excelência o governador-geral de Angola.* Lisbon: E.N.P, 1970.

_____. *Estatutos da Companhia de Diamantes de Angola.* Lisbon: Neogravura, 1939.

_____. *Estatutos da Companhia de Diamantes de Angola.* Lisbon: Oficina Grafica, 1950.

_____. *Estatutos e Diplomas de Concessão da Companhia de Diamantes de Angola.* Lisbon: Companhia de Diamantes de Angola, 1971.

_____. *Extracto de entrevista com o ex-chefe da zôna mineira da unda, engenheiro sr. Eugenio Salles Lane, publicada no jornal "A Patria," nos. 53, 54 e 55 de 24 e 27 de Novembro e 1 de Dezembro de 1925.* Luanda: Empreza Gráfica de Angola, 1925.

_____. *Legislação penal referente à exploração de diamantes na colónia de Angola.* Lisbon: Oficina Gráfica, 1948.

_____. *Legislação penal respeitante à exploração de diamantes na província de Angola e disposições contratuais com ela conexas.* Lisbon: Tip. Silvas, 1968.

_____. *Notícia succinta sobre a sua constituição, concessões obtidas e trabalhos realizados em Angola.* Lisbon: Diamang, 1929.

_____. *Relatório de conselho de administração.* Luanda: Diamang, 1953.

_____. *O romance do luachimo.* Lisbon: Bertrand, 1969.

_____. *Ordem de serviço (circular) no. 3-D/48: Instruções gerais sobre mão de obra indigena.* January 26, 1948.

_____. *Perspectiva de acção social.* Lisbon: Neogravura, 1973.

_____. *Relatório e projecto de cisão dos bens situados em Portugal que pertenceram à Companhia de Diamantes de Angola, S. A. R. L.* Lisbon: C. D. A, 1979.

_____. *Suplemento à compilação dos estatutos e diplomas de concessão da companhia de diamantes de Angola, contendo novos frontispicio e indice, estatutos actualizados decreto-lei n'39.920 e contrato celebrado com o governo em 10 de fevereiro de 1955.* Lisbon: Tip. Casa Portuguesa, 1955.

_____. *Tráfico ilegal de diamantes: O que é, por que existe, quais as suas consequências económicas e sociais para Angola (Excerto de entrevista publicada em "A província de Angola de 12 e 14 de Abril de 1972, concedida pelo presidente do conselho de administração da DIAMANG, Eng. Carlos Krus Abecasis).* Luanda: Diamang, 1973.

Diamang Direcção-Geral na Lunda, *Relatório no. A-549* (December 1961).

Diamang, Sociedade Anonima de Responsabilidade Limitada com o Capital de 3.000.000$00. *Relatório do conselho de administração e parecer do conselho fiscal.* Lisbon: Tipografia de Papelaria de Moda, 1920–21.

Diamang, Sociedade Anonima de Responsabilidade Limitada com o Capital de 9.000.000$00. *Relatório do conselho de administração e parecer do conselho fiscal.* Lisbon: 1925, 1928–29, 1931–38, 1940–44.

Diamang, Sociedade Anónima de Responsibilidade Limitada com o Capital de 179.300.000$00. *Relatório de conselho de administração e parecer do conselho fiscal.* Lisbon: 1945–54.

Diamang, Sociedade Anónima de Responsabilidade Limitada com o capital de 294,100,100$00. *Relatório do conselho de administração e parecer do conselho fiscal.* Lisbon: 1955–67.

Dias, Jill R., Rosa Cruz e Silva, Ana Oliveira, et al. *Actas do II seminário internacional sobre a história de Angola: Construindo o passado Angolano: As fontes e a sua interpretação. Luanda, 4 a 9 de Agosto de 1997.* Lisbon: CNCDP, 2000.

Dilwa, Carlos Rocha. *Contribuição à história económica de Angola.* Luanda: I.N.A, 1978.

Direcção do Serviços da Administração Civil de Angola. *Código do trabalho dos indígenas.* Luanda: Imprensa Nacional de Angola, 1936.

Duffy, James. *Portuguese Africa.* Cambridge, MA: Harvard University Press, 1961.

Endiama. *Answers for You.* Luanda: Endiama, 2004.

———. *Razões para financiar o sector diamantífero de Angola.* Luanda: Endiama, 2004.

Esteves, Emmanuel. "O caminho-de-ferro de benguela e o impacto económico, social e cultural na sua zona de influência (1902–1952)." *Africana Studia* 3 (2000): 49–72.

Federação Regional dos Sindicatos dos Empregados de Escritório do Sul e Ilhas Adjacentes. *Acordo colectivo de trabalho do pessoal ao serviço da Companhia de Diamantes de Angola.* Lisbon: Fed. Reg. dos Sind. dos Empregados de Escritório do Sul e Ilhas Adjacentes, 1975.

Ferguson, James. *Global Shadows: Africa in the Neoliberal World Order.* Durham, NC: Duke University Press, 2006.

Fetter, Bruce. *The Creation of Elisabethville, 1910–1940.* Stanford, CA: Hoover Institution Press, 1976.

Freitas, João Araújo de. "Trabalhadores indígenas de Angola, sua alimentação, doenças predominantes e algumas medidas profilácticas adoptadas." *Seperata dos Anais do Instituto de Medicina Tropical* 10, no. 3 (September 1953): 1157–84.

Freund, Bill. *The African Worker.* Cambridge: Cambridge University Press, 1988.

Freyre, Gilberto. *Aventura e rotina: Sugestões de uma viagem à procura das constantes portuguesas de caráter e acção.* Rio de Janeiro: Livaria José Olympio Editoria, 1953.

Galvão, Henrique. *Exposição do deputado Henrique Galvão, à comissão de colónias da Assembleia Nacional, em Janeiro de 1947.* Lisbon: Arquivo Histórico-Parlamentar, Assembleia da República, 1947.

_____. *Informação económica sôbre Angola*. Lisbon: Ed. Direcção das Feiras de Amostras Coloniais, 1932.

_____. *Outras terras, outras gentes: 25,000 quilómetros em Angola*. vol. 1. Lisbon: Tipografia Silvas, 1942.

_____. *Por Angola: Quatro anos de actividade parliamentar*. Lisbon: Edição do Autor, 1949.

_____. *Santa Maria: My Crusade for Portugal*. Cleveland, OH: World, 1961.

Gengenbach, Heidi. "Boundaries of Beauty: Tattooed Secrets of Women's History in Magude District, Southern Mozambique." *Journal of Women's History* 14, no. 4 (2003): 106–41. http://dx.doi.org/10.1353/jowh.2003.0007.

Gluckman, Max. "Tribalism in Modern British Central Africa." In *Africa: Social Problems of Change and Conflict*, edited by Pierre L. Van Den Berghe, 55–70. San Francisco: Chandler, 1965.

Gomez, Michael A. *Exchanging Our Country Marks: The Transformation of African Identities in the Colonial and Antebellum South*. Chapel Hill: University of North Carolina Press, 1998.

Gordon, Christine, ed. *Diamond Industry Annual Review: Republic of Angola 2005*. Ottawa: Partnership Africa Canada, 2005.

Gordon, Robert J. *Mines, Masters and Migrants: Life in a Namibian Compound*. Johannesburg: Ravan Press, 1977.

Grier, Beverly. "Invisible Hands: The Political Economy of Child Labor in Colonial Zimbabwe, 1890–1930." *Journal of Southern African Studies* 20, no. 1 (March 1994): 27–52. http://dx.doi.org/10.1080/03057079408708385.

Hailey, Lord William Malcolm. *An African Survey: A Study of Problems Arising in Africa South of the Sahara*. London: Oxford University Press, 1938.

Harries, Patrick. *Work, Culture, and Identity: Migrant Laborers in Mozambique and South Africa, c. 1860–1910*. Portsmouth, NH: Heinemann, 1994.

Henriques, Isabel. *Território e identidade: A construção da Angola colonial, c. 1872–c. 1926*. Lisbon: Universidade de Lisboa, 2004.

Heywood, Linda. *Contested Power in Angola, 1840s to the Present*. Rochester, NY: University of Rochester Press, 2000.

_____. "Slavery and Forced Labor in the Changing Political Economy of Central Angola, 1850–1949." In *The End of Slavery in Africa*, edited by Suzanne Miers and Richard Roberts, 415–36. Madison: University of Wisconsin Press, 1988.

Higginson, John. *A Working Class in the Making: Belgian Colonial Labor Policy, Private Enterprise, and the African Mineworker, 1907–1951*. Madison: University of Wisconsin Press, 1989.

Huibregtse, P. K. *Angola, The Real Story*. Translated by Nicolette Buhr. The Haag: Zuid-Hollandsche, 1973.

Hunt, Nancy Rose. *A Colonial Lexicon: Of Birth Ritual, Medicalization and Mobility in the Congo*. Durham, NC: Duke University Press, 1999. http://dx.doi.org/10.1215/9780822381365.

International Labour Organization (ILO). *International Labor Conference, 56th Session, Geneva 1971. Report III Part 4A. Report of the Committee of Experts on the Application of Conventions and Recommendations.*

ILO. International Labor Conference, 56th Session, Geneva 1971. Report III Part 4A. *Report of the Committee of Experts on the Application of Conventions and Recommendations.* Report on Direct Contacts with the Government of Portugal regarding the Implementation of the Abolition of Forced Labour Convention 1957 (no. 105). Report by Pierre Juvigny, Representative of the Director-General of the ILO, on Direct Contacts with the Government of Portugal Regarding the Implementation of the Abolition of Forced Labour Convention 1957 (no. 105).

ILO Official Bulletin, # 45 (XLV) 1962. Supplement II, No. 2 April 1962. Report of the Commission Appointed under Article 26 of the Constitution of the ILO to Examine the Complaint Filed by the Government of Ghana concerning the Observance by the Government of Portugal of the Abolition of Forced Labor Convention, 1957 (No. 105).

ILO. Report of the Ad Hoc Committee on Forced Labor, ILO. New Series of the ILO, Geneva, 1953. Memorandum of February 22, 1952.

International Monetary Fund. *Angola: Selected Issues and Statistical Appendix 2005, April 2005.* IMF Country Report No. 05/125, 2005.

Isaacman, Allen. "Coercion, Paternalism and the Labour Process: The Mozambican Cotton Regime, 1938–1961." *Journal of Southern African Studies* 18, no. 3 (September 1992): 487–526.

———. *Cotton Is the Mother of Poverty: Peasants, Work, and Rural Struggle in Colonial Mozambique, 1938–1961.* Portsmouth, NH: Heinemann, 1996.

———. "Displaced People, Displaced Energy, and Displaced Memories: The Case of Cahora Bassa, 1970–2004." *International Journal of African Historical Studies* 38, no. 2 (2005): 201–38.

Isaacman, Allen, and Barbara S. Isaacman. *Slavery and Beyond: The Making of Men and Chikunda Ethnic Identities in the Unstable World of South-Central Africa, 1750–1920.* Portsmouth, NH: Heinemann, 2004.

James, Wilmot G. *Our Precious Metal: African Labour in South Africa's Gold Industry, 1970–1990.* Bloomington: Indiana University Press, 1992.

Joyce, Donald Franklin. *Slavery in Portuguese Africa: Opposing Views (includes Henry Wood Nevinson's "A Modern Slavery" and Francisco Mantero's "Portuguese Planters and British Humanitarians: The Case for S. Thomé").* Translated by Lieut. Col. J. A. Wyllie. Northbrook, IL: Metro Books, 1972.

Júnior, João Vieira Santa Ana. *Diamantes de Angola: Um grave problema nacional.* Vol. 2, part 1a. (n.d.).

———. *Factos verídicos sobre a exploração diamantífera de Angola.* Lisbon: LitoGráfica, 1969.

———. *O caso Santa Ana Júnior—Diamang, condiama.* Coimbra: Gráfica de Coimbra, 1973.

Kreike, Emmanuel. *Re-creating Eden: Land Use, Environment, and Society in Southern Angola and Northern Namibia.* Portsmouth, NH: Heinemann, 2004.

Leal, Cunha. *Coisas do tempo presente, I: Coisas da Companhia de Diamantes de Angola (Diamang)*. Lisbon: Edição do Autor, 1957.

———. *Coisas do tempo presente, II: Novas coisas da Companhia de Diamants de Angola (Diamang)*. Lisbon: Edição do Autor, 1959.

———. *Coisas do tempo presente: Peregrinações através do poder económico*. Lisbon: Edição do Autor, 1960.

Lindsay, Lisa A. *Working with Gender: Wage Labor and Social Change in Southwestern Nigeria*. Portsmouth, NH: Heinemann, 2003.

Lourenço, Dr. José Pires: *A exploração dos diamantes em Angola*. Lisbon: Editorial Império, 1957.

Marcum, John A. *The Angolan Revolution: The Anatomy of an Explosion (1950–1962)*. Vol. 1. Cambridge, MA: MIT Press, 1969.

Marques, Rafael. *Angola's Deadly Diamonds: Lundas, the Stones of Death*. 2005.

———. *Diamantes de sangue: Corrupção e tortura em Angola*. Lisbon: Tinta da China, 2011.

———. *Operação kissonde: Os diamantes da humilhação e da miséria*. 2006.

Martins, João Vicente. *Crenças, adivinhação e medicina tradicionais dos Tutchokwe do nordeste de Angola*. Lisbon: Edição do Instituto de Investigação Ciêntifica Tropical, 1993.

Martins, Silva. Relatório "Assistência Sanitária—1950." Dundo: October 16, 1950.

Mastrobuono, Luisa. "Ovimbundu Women and Coercive Labour Systems, 1850-1940: From Still Life to Moving Picture." MA thesis, University of Toronto, 1992.

Matos, Albuquerque. "Comunicações." In *Diamante*, edited by Bernardo Reis, XXI encontro Diamang, manique do intendente, 14 June 2003. Braga (2003): 15.

McCormick, Shawn. *The Angolan Economy: Prospects for Growth in a Postwar Environment*. Washington, DC: Center for Strategic and International Studies, 1994.

McNamara, J. K. "Brothers and Work Mates: Home Friend Networks in the Social Life of Black Migrant Workers in a Gold Mine Hostel." In *Black Villagers in an Industrial Society*, edited by Philip Mayer, 305–53. Cape Town: Oxford University Press, 1980.

Mendes, Afonso. *A huíla e moçâmedes: Considerações sobre o trabalho indígena*. Lisbon: Tipografia Minerva, 1958.

———. "Breves considerações sobre mão-de-obra Africana." *Ultramar* 4 (1961): 1–15.

———. *O trabalho assalariado em Angola*. Lisbon: Instituto Superior de Ciências Sociais e Política Ultramarina, 1966.

———. *Serviços de colocação, separata do boletim "Trabalho,"* 1–19. Luanda: Instituto do Trabalho, Previdência e Acção Social, 1963.

———. *Tribunais de trabalho, separata do boletim "Trabalho,"* 1–24. Luanda: Instituto do Trabalho, Previdência e Acção Social, 1963.

Milheiros, Mário. *Índice histórico-corográfico de Angola*. Luanda: Instituto de Investigação Científica de Angola, 1972.

Miller, Joseph C. *Cokwe Expansion, 1850–1900*. Madison: African Studies Program, University of Wisconsin, 1969.

———. "Chokwe Trade and Conquest in the Nineteenth Century." In *Pre-colonial African Trade: Essays on Trade in Central and Eastern Africa before 1900*, edited by Richard Gray and David Birmingham, 175–201. London: Oxford University Press, 1970.

Minter, William, ed. *Operation Timber: Pages from the Savimbi Dossier*. Trenton, NJ.: Africa World Press, 1988.

Moodie, Dunbar. "Mine Culture and Miners' Identity on the South African Gold Mines." In *Town and Countryside in the Transvaal: Capitalist Penetration and Popular Response*, edited by Belinda Bozzoli, 176–97. Johannesburg: Ravan Press, 1983.

Moodie, T. Dunbar. "Ethnic Violence on South African Gold Mines." *Journal of Southern African Studies* 18, no. 3 (1992): 584–613. http://dx.doi.org/10.1080/03057079208708327.

———. "The Formal and Informal Social Structure of a South African Gold Mine." *Human Relations* 33, no. 8 (1980): 555–74. http://dx.doi.org/10.1177/001872678003300803.

Moodie, T. Dunbar, and Vivienne Ndatshe. *Going for Gold: Men, Mines, and Migration*. Berkeley: University of California Press, 1994.

Moorman, Marissa. "'Feel Angolan with this Music': A Social History of Music and Nation, Luanda, Angola, 1945–1975." PhD diss., University of Minnesota, 2004.

———. *Intonations: A Social History of Music and Nation in Luanda, Angola, from 1945 to Recent Times*. Athens: Ohio University Press, 2008.

Nafisi, Azar. *Reading Lolita in Tehran: A Memoir in Books*. New York: Random House, 2003.

Nascimento, Hermenegildo A. "África e os diamantes." *"Endiama Hoje" Revista Informativa de Empresa nacional de Diamantes de Angola* (July–August 2004): 32.

———. "Kimberlitos e lamproitos: Rochas abençoadas." *"Endiama Hoje" Revista Informativa de Empresa nacional de Diamantes de Angola* (January–February 2004): 25.

N'Diaye, Boubacar. "Ivory Coast's Civilian Control Strategies, 1961–1998: A Critical Assessment," *Journal of Political and Military Sociology* 28, no. 2 (2000): 246–70.

Neto, Ana Maria. *Industrialização de Angola: Reflexão sobre a experiência da administração portuguesa (1961–1975)*. Lisbon: Escher, 1991.

Nevinson, Henry W. *A Modern Slavery*. New York: Harper and Brothers, 1906.

Newitt, Malyn. *Portugal in Africa: The Last Hundred Years*. Harlow, Essex: Longman, 1981.

Okuma, Thomas. *Angola in Ferment: The Background and Prospects of Angolan Nationalism*. Boston: Beacon Press, 1962.

Oliveira, José Carlos de. *O comerciante do mato: O comércio no interior de Angola e Congo*. Coimbra: Centro de Estudos Africanos, 2004.

Parkinson, Lute J. *Memoirs of African Mining*. Self-published, 1962.

Parpart, Jane L. "The Household and the Mine Shaft: Gender and Class Struggles on the Zambian Copperbelt, 1926–64." *Journal of Southern African Studies* 13, no. 1 (1986): 36–56. http://dx.doi.org/10.1080/03057078608708131.

———. *Labor and Capital on the African Copperbelt*. Philadelphia: Temple University Press, 1983.

———. "'Wicked Women' and 'Respectable Ladies': Reconfiguring Gender on the Zambian Copperbelt, 1936–1964." In *Wicked Women and the Reconfiguration of Gender in Africa*, edited by Dorothy L. Hodgson and Sheryl A. McCurdy, 274–92. Portsmouth, NH: Heinemann, 2001.

Pélissier, René. *História das campanhas de Angola: Resistência e revoltas, 1845–1941*. 2 vols. Lisbon: Editorial Estampa, 1986.

Penvenne, Jeanne Marie. *African Workers and Colonial Racism: Mozambican Strategies and Struggles in Lourenço Marques, 1877–1962*. Portsmouth, NH: Heinemann, 1995.

Perrings, Charles. *Black Mineworkers in Central Africa: Industrial Strategies and the Evolution of an African Proletariat in the Copperbelt, 1911–41*. New York: Africana, 1979.

———. "'Good Lawyers but Poor Workers': Recruited Angolan Labour in the Copper Mines of Katanga, 1917–1921." *Journal of African History* 18, no. 2 (1977): 237–59. http://dx.doi.org/10.1017/S0021853700015516.

Phimister, Ian R., and Charles Van Onselen. *Studies in the History of African Mine Labour in Colonial Zimbabwe*. Gwelo: Mambo Press, 1978.

Picoto, José. *Assistência médico-cirúrgica na Lunda pelo serviço de saúde da Diamang: Elementos estatísticos de cinco anos de actividade (Separata dos anais do Instituto de Medicina Tropical, Volume X, No. 4, Fasc. I, Setembro de 1953)*. Porto: Imprensa Portuguesa, 1953.

———. *Uma acção médico-sanitária em África altamente dignificante para Portugal: Assistência médico-sanitária na Lunda pela Companhia de Diamantes de Angola*. Porto: Costa Carregal, 1963.

Pitcher, M. Anne. "From Coercion to Incentives: The Portuguese Colonial Cotton Regime in Angola and Mozambique." In *Cotton, Colonialism, and Social History in Sub-Saharan Africa*, edited by Allen Isaacman and Richard Roberts, 119–43. Portsmouth, NH: Heinemann, 1995.

———. "Sowing the Seeds of Failure: Early Portuguese Cotton Cultivation in Angola and Mozambique, 1820–1926." *Journal of Southern African Studies* 17, no. 1 (March 1991): 43–70. http://dx.doi.org/10.1080/03057079108708266.

Pongweni, Alec J. C. "The Chimurenga Songs of the Zimbabwean War of Liberation." In *Readings in African Popular Culture*, edited by Karin Barber, 63–72. Bloomington: University of Indiana Press, 1997.

Porto, Nuno. *Angola a preto e branco: Fotografia e ciência no museu do Dondo, 1940–1970*. Coimbra: Museu Antropológico da Universidade de Coimbra, 1999.

_____. "Manageable Past: Time and Native Culture at the Dundo Museum in Colonial Angola." *Cahiers d'Études Africaines* 39, no. 155 (1999): 767–87. http://dx.doi.org/10.3406/cea.1999.1777.

_____. "'Under the Gaze of Ancestors': Photographs and Performance in Colonial Angola." In *Photographs, Objects, Histories*, edited by Elizabeth Edwards and Janice Hart, 113–31. London: Routledge, 2004.

Ramphele, Mamphela. *A Bed Called Home: Life in the Migrant Labour Hostels*. Athens: Ohio University Press, 1993.

Redinha, José. *Distribuição étnica da província de Angola*. Luanda: Centro de Informação e Turismo de Angola, 1970.

Reis, Bernardo, ed. *Diamante*, XXI encontro Diamang, manique do intendente, 14 June 2003. Braga, 2003.

_____. *Diamante*, XXII encontro Diamang, tomar, 12 June 2004. Braga, 2004.

Reis, Bernardo, and Vítor Sousa, eds. *Diamante*. XVII encontro Diamang. Braga, 1999.

República Portuguesa. *Província de Angola: Regulamento ao código do trabalho dos indígenas, aprovado pelo diploma legislativo no. 2:797 de 31 de Dezembro de 1956*. Luanda: Imprensa Nacional de Angola, 1957.

Ribeiro, Orlando. *A colonização de Angola e o seu fracasso*. Lisbon: Impressa Nacional, 1981.

Ross, Edward Alsworth. *Report on Employment of Native Labor in Portuguese Africa*. New York: Abbott Press, 1925.

Rukanshagiza, Joseph B. "African Armies: Understanding the Origin and Continuation of Their Non-Professionalism." PhD diss., State University of New York, Albany, 1995.

Sá, Vasco de. *A Lunda . . . os diamantes . . . a endiama*. Luanda: ELO—Publicidade, Artes Gráficas, 1996.

_____. *Endiama: Uma empresa diferente*. Luanda: Elo-Publicidade, Artes Gráficas, 1997.

_____, interview with. "No começo foi aventura." *"Endiama Hoje" Revista Informativa de Empresa nacional de Diamantes de Angola* Ano 3 (January–February 2004: 19–21.

Schmidt, Elizabeth. *Peasants, Traders and Wives: Shona Women in the History of Zimbabwe, 1870–1939*. Portsmouth, NH: Heinemann, 1992.

Scott, James C. *Seeing Like a State: How Certain Schemes to Improve the Human Condition Have Failed*. New Haven, CT: Yale University Press, 1998.

_____. *Weapons of the Weak: Everyday Forms of Peasant Resistance*. New Haven, CT: Yale University Press, 1985.

Silva, Cunha. *O trabalho indígena: Estudo de direito colonial*. Lisbon: Agência Geral do Ultramar, 1955.

Sociedade Portuguesa de Lapidação de Diamantes. *Estatutos da sociedade portuguesa de lapidação de diamantes*. Lisbon: Silvas, 1965.

Soromenho, Castro. *Terra morta*. Rio de Janeiro: Casa do Estudante do Brasil, 1949.

_____. *Sertanejos de Angola*. Lisbon: Agência Geral das Colónias, 1943.

Tavares, Ana Paula Ribeiro. "História e Memória: Estudo sobre as Sociedades Lunda e Cokwe de Angola." PhD diss., Universidade Nova de Lisboa, 2009.

Teixeira, Alberto de Almeida. *Lunda: Sua organização e ocupação*. Lisbon: Agência Geral dos Colónias, 1948.

Turrell, Robert. *Capital and Labour on the Kimberley Diamond Fields, 1871–1890*. Cambridge: Cambridge University Press, 1987.

_____. "Diamonds and Migrant Labor in South Africa, 1869–1910." *History Today* 36 (May 1986): 45–49.

_____. "Kimberley's Model Compounds." *Journal of African History* 25, no. 1 (1984): 59–75. http://dx.doi.org/10.1017/S0021853700022568.

Vail, Leroy, and Landeg White. *Capitalism and Colonialism in Mozambique: A Study of Quelimane District*. London: Heinemann, 1980.

_____. "Plantation Protest: The History of a Mozambican Song." In *Readings in African Popular Culture*, edited by Karin Barber, 54–63. Bloomington: University of Indiana Press, 1997.

Valente, Antunes. "Assistência médica aos trabalhadores em Angola." *Seperata do Jornal do Médico* 70 (November 1969): 1–29.

Van Onselen, Charles. "Black Workers in Central African Industry: A Critical Essay on the Historiography and Sociology of Rhodesia." *Journal of Southern African Studies* 1, no. 2 (April 1975): 228–46. http://dx.doi.org/10.1080/03057077508707935.

_____. *Chibaro: African Mine Labour in Southern Rhodesia, 1900–1933*. London: Pluto Press, 1976.

_____. *The Seed Is Mine: The Life of Kas Maine, a South African Sharecropper, 1894–1985*. New York: Hill and Wang, 1996.

Varanda, Jorge. "'A Bem da Nação': Medical Science in a Diamond Company in Twentieth-Century Colonial Angola." PhD diss., University College, London, 2007.

Veatch, Arthur Clifford. *Evolution of the Congo Basin*. Washington, DC: Judd and Detweiler, 1935.

Vellut, Jean-Luc. "Mining in the Belgian Congo." In *History of Central Africa*, vol. 2, edited by David Birmingham and Phyllis M. Martin, 126–62. London: Longman, 1983.

Vilaça, Alberto de Oliveira. *Diamantes de Angola mas não para Angola*. Coimbra: Agueda, 1972.

Watângua, Manuel. "Desenvolvimento deactividades Geológico-mineiras em aluviões." *"Endiama Hoje" Revista Informativa de Empresa nacional de Diamantes de Angola* (July–August 2004): 23–24.

Wheeler, Douglas L. "The Forced Labor 'System' in Angola, 1903–1947: Reassessing Origins and Persistence in the Context of Colonial Consolidation, Economic Growth and Reform Failures." Presented: November 18, 2005 at the II Coloquio Internacional "Trabalho Forcado Africano", Centro de Estudos Africanos, Univ. do Porto, Portugal.

_____. "José Norton de Matos (1867–1955)." In *African Proconsuls: European Governors in Africa*, edited by L. H. Gann and Peter Duignan, 445–66. New York: Free Press, 1978.

Wheeler, Douglas L., and René Pélissier. *Angola*. New York: Praeger, 1971.

White, Landeg. *Magomero: Portrait of an African Village*. Cambridge: Cambridge University Press, 1987.

Worger, William H. *South Africa's City of Diamonds: Mine Workers and Monopoly Capitalism in Kimberley, 1867–1895*. New Haven, CT: Yale University Press, 1987.

Index

www.ingramcontent.com/pod-product-compliance
Lightning Source LLC
Chambersburg PA
CBHW072054020426
42334CB00017B/1502